The Left-handed Spinor

Chiral Algebras

The Anti-matter and Matter Imbalance in the Universe

Parity and CP Invariance in the Universe

The Violation of Parity
and the
Violation of CP Invariance
in the Universe

By

Dennis Morris

Contents

Contents

Contents

Contents

Contents

Introduction

This book is a sequel to the very successful book 'The Naked Spinor' by the same author. In 'The Naked Spinor' we saw that Clifford algebras and spinors are just division algebras which derive from the $C_2 \times C_2 \times ...$ finite groups. In this book, we see that there are chiral spinors. By this we mean that there are left-handed spinors and right-handed spinors. This means there are left-handed division algebras and right-handed division algebras – chiral algebras. These chiral algebras have considerable potential as fields within particle physics, and it is becoming apparent that the neutrino field is a left-handed spinor field.

In this book, we explore the chiral algebras. Rather shockingly, we discover that there are only five types of chiral algebra which can be manifest in our 4-dimensional space-time. There are two 4-dimensional chiral algebras and three 8-dimensional chiral algebras which can be manifest in our 4-dimensional space-time. It seems that these five algebras, together with the C_2 algebras, comprise the whole of our universe. The two 4-dimensional chiral algebras seem to be our 4-dimensional space-time, the classical forces of gravity and classical electromagnetism, and the weak force of the electron and the neutrino. The three 8-dimensional chiral algebras seem to be the strong force associated with quarks.

We discover parity within these chiral algebras, and we discover violation of parity within these chiral algebras. We discover charge conjugation within these chiral algebras, and we discover violation of charge conjugation within these chiral algebras. We discover CP invariance within these chiral algebras, and we discover violation of CP invariance within these chiral algebras. All of this fits with observations of the physics of our universe made by particle physicists.

We discover anti-matter within these chiral algebras, and we discover a matter/anti-matter imbalance within these chiral algebras which seems to fit with observations of our universe. We come to understand the complicated nature of anti-matter which appears in many guises depending upon the particle involved. The strange behaviour of the neutrino and of the kaon and B-mesons seems to be within these chiral algebras.

Prior to the publishing of this book, the chirality of algebras was unknown. The discovery of algebraic chirality will doubtless be a landmark in the history of both mathematics and theoretical physics.

Apologies:

Allow me to apologise for the uncomfortable size of this book. I would have preferred to produce a smaller and more easily handled book, but the tables of data would simply not fit on to a physically smaller page. Also allow me to apologise for the occasional repetition and reiteration within this book; it is done with the intention of making this book easier to read and easier to understand.

I must also apologise for the incompleteness of this book. I would have liked to have presented the whole of theoretical physics neatly tied up with no loose ends and to have declared that we now understand everything about the chiral algebras and their involvement in physics. We do not fully understand the chiral algebras, and our understanding of the place of the chiral algebras in physics is still rudimentary. There is still much work to be done before we can claim to properly understand these chiral algebras;

this is particularly true of the 8-dimensional chiral algebras of which there are one thousand and twenty four in three algebraically non-isomorphic types.

In this book, we cover the 16-dimensional $C_2 \times C_2 \times C_2 \times C_2$ algebras in only one short chapter. I apologise for this shallowness of presentation of the 16-dimensional $C_2 \times C_2 \times C_2 \times C_2$ algebras, but one short chapter is all we need to show that the 16-dimensional $C_2 \times C_2 \times C_2 \times C_2$ algebras are not chiral algebras. This does not mean the 16-dimensional algebras are of no interest, and here there is much virgin territory to be explored. A book of this physical size cannot hold the 16×16 matrices needed to properly present the 16-dimensional $C_2 \times C_2 \times C_2 \times C_2$ algebras, and, as I have said above, the physical size of this book is already uncomfortable.

I hope the reader will find this book clear and easy to read even though the material covered is complicated. I hope the reader will take this knowledge forward into their own understanding of the universe. Having read this far, the reader already knows that some algebras are left-handed and other algebras are right-handed. This, in itself, is quite a profound change to our understanding of mathematics.

Dennis Morris

Brotton

February 2017

A Notational Note

We write the complex numbers, $\mathbb{C}$, in matrix form:

$$a + ib = \begin{bmatrix} a & b \\ -b & a \end{bmatrix} \tag{0.1}$$

Similarly, we write the quaternions in matrix form:

$$\mathbb{H}_{L\chi} = \begin{bmatrix} a & b & c & d \\ -b & a & -d & c \\ -c & d & a & -b \\ -d & -c & b & a \end{bmatrix} \qquad \mathbb{H}_{R\chi} = \begin{bmatrix} a & b & c & d \\ -b & a & d & -c \\ -c & -d & a & b \\ -d & c & -b & a \end{bmatrix} \tag{0.2}$$

We similarly write all division algebras in matrix form.

There are many advantages to the matrix notation; we list a few:

1) The matrix form of a division algebra clearly displays the particular finite group, presented as permutation matrices, which underlies the particular division algebra.
2) The matrix form of a division algebra allows the simple calculation, by taking the matrix exponential of the matrix form, of the rotation matrix of the particular division algebra and of the trigonometric functions of the particular division algebra.
3) The matrix form of a division algebra allows the calculation shortcut of differentiation by the simple application of a differential operator; this saves a lot of paper and a lot of time. We point out that, really, in algebra, there is no such thing as a differential operator; it is just a calculation aid.
4) The matrix form of a division algebra allows the presentation of the different matrix forms of an algebra, see above, (0.2); this is almost impossible using non-matrix notation. Without the matrix form of a division algebra, we would be unaware of chirality in mathematics, as, indeed, we were prior to the use of the matrix form notation of a division algebra. Without the matrix form notation of a division algebra, this book would not have been written.
5) We see that the multiplication operation within a division algebra, like the complex numbers or the quaternions, is just matrix multiplication. Of course, matrix multiplication within a division algebra is just a 'beefed up' version of the sequential combination of permutations within a finite group[1].

---

[1] See : Dennis Morris : Finite Groups A Simple Introduction.

Chapter 1

Algebraic Chirality

Multiplicative commutativity:

There are multiplicatively commutative division algebras like the 2-dimensional complex numbers, $\mathbb{C}$. In a multiplicatively commutative algebra, the order in which we multiply two elements of the algebra together does not change the product of the multiplication. If $B \& C$ are two elements of a commutative algebra, then we have $BC = CB$. Indeed this is exactly what we mean by multiplicative commutativity. Multiplicative commutativity can be expressed using the anti-commutator and the commutator:

$$
\begin{array}{cc}
Anti-commutator & Commutator \\
\{B \quad C\} = BC + CB = 2BC & [B \quad C] = BC - CB = 0
\end{array}
\tag{1.1}
$$

The commutator is zero for every pair of elements in a commutative algebra. The anti-commutator is non-zero for every pair of non-zero elements in a commutative algebra.

Multiplicative non-commutativity:

Within finite group theory, we find non-commutative finite groups. Within a non-commutative group, the order in which we multiply two elements of the group together most often does change the product of the multiplication. Non-commutative groups have some elements which are such that $BC = D$ and $CB = E \neq \pm D$. It is inappropriate to express this kind of finite group non-commutativity using the anti-commutator and the commutator; if we try to do this, we get:

$$
\{B \quad C\} = BC + CB = D + E \qquad\qquad [B \quad C] = BC - CB = D - E
\tag{1.2}
$$

Within a finite group, there is only one operation, which we call multiplication, and so the plus sign and the minus sign in (1.2) are meaningless.

There are division algebras which derive from every finite group[2], including the non-commutative finite groups. Within division algebras, we have two algebraic operations, addition and multiplication, and so, within division algebras, the commutator and anti-commutator do make sense; the anti-commutator and the commutator are not very useful in division algebras derived from the non-commutative finite groups, but they do make sense.

Algebras from a non-commutative group:

We give an example of the division algebras which arise from a non-commutative finite group. The order six finite group S_3 is a non-commutative group. The separate algebraic matrix forms arise by taking the

---

[2] See: Dennis Morris : Complex Numbers The Higher Dimensional Forms.

4

permutations $P_{i,j} = \pm 1$ of the parameters within the general algebraic matrix form of this group[3]. The individual division algebras are then derived by taking the matrix exponential of each algebraic matrix form[4] to produce the polar form of the algebra. There are $2^5 = 32$ separate division algebras within the group S_3 because this is the number of possible permutations of the five parameters $P_{i,j} = \pm 1$. These thirty-two algebras are of six algebraically non-isomorphic types.[5] The general algebraic matrix form of the S_3 group is:

$$S_3 =$$

$$
\begin{bmatrix}
a & b & c & d & e & f \\[2ex]
P_{2,1}c & a & \dfrac{(P_{2,1})^2 (P_{2,4})^3}{(P_{3,4})^3 (P_{5,6})^3}b & P_{2,4}e & \dfrac{(P_{2,1})^2 (P_{2,4})^2}{(P_{3,4})^2 (P_{5,6})^3}f & \dfrac{P_{2,1}}{P_{3,4}}d \\[3ex]
P_{2,1}b & \dfrac{(P_{3,4})^3 (P_{5,6})^3}{P_{2,1}(P_{2,4})^3}c & a & P_{3,4}f & \dfrac{P_{2,1}}{P_{2,4}}d & \dfrac{(P_{3,4})^3 (P_{5,6})^3}{P_{2,1}(P_{2,4})^2}e \\[3ex]
P_{4,1}d & \dfrac{(P_{3,4})^2 P_{4,1} (P_{5,6})^2}{(P_{2,1})^2 P_{2,4}}e & \dfrac{(P_{2,4})^2 P_{4,1}}{P_{3,4}(P_{5,6})^2}f & a & \dfrac{(P_{2,1})^2 P_{2,4}}{(P_{3,4})^2 (P_{5,6})^2}b & \dfrac{P_{3,4}(P_{5,6})^2}{(P_{2,4})^2}c \\[3ex]
\dfrac{(P_{3,4})^2 P_{4,1}(P_{5,6})^2}{(P_{2,1})^2}e & \dfrac{(P_{3,4})^2 P_{4,1}P_{5,6}}{(P_{2,1})^2}f & \dfrac{P_{2,4}P_{4,1}}{P_{2,1}}d & \dfrac{(P_{3,4})^2 (P_{5,6})^2}{P_{2,1}P_{2,4}}c & a & P_{5,6}b \\[3ex]
\dfrac{(P_{2,4})^2 P_{4,1}}{(P_{5,6})^2}f & \dfrac{P_{4,1}P_{3,4}}{P_{2,1}}d & \dfrac{(P_{2,4})^2 P_{4,1}}{P_{2,1}P_{5,6}}e & \dfrac{P_{2,1}(P_{2,4})^2}{P_{3,4}(P_{5,6})^2}b & \dfrac{P_{2,1}}{P_{5,6}}c & a
\end{bmatrix}
$$

$$(1.3)$$

We can separate the above algebraic matrix form, (1.3), into six matrices each of which holds one of the variables $\{a,b,c,d,e,f\}$. We denote these six variable matrices by $\{A,B,C,D,E,F\}$ corresponding to the six variables. Each of the variable matrices corresponds to an element of the group S_3; A is the identity element etc..

Taking all parameters, $P_{i,j} = +1$ in (1.3), we find:

$$EF = B \quad \& \quad FE - C \tag{1.4}$$

Changing the values of the parameters, $P_{i,j}$, will change nothing except the signs in (1.4); it will not change the variables. In the case of the algebra with all parameters, $P_{i,j} = +1$, the anti-commutator and the commutator of the $E \& F$ variable matrices are:

$$\{E \quad F\} = EF + FE = B + C \qquad\qquad [E \quad F] = EF - FE = B - C \tag{1.5}$$

---

[3] Although we usually take the parameters to be plus unity or minus unity, they can take any non-zero value. It is the signs that are important.

[4] Only the polar form is a division algebra. The individual algebraic matrix forms have singular matrices and zero divisors of which we are rid when we take the matrix exponential.

[5] See: Dennis Morris : The Uniqueness of our Space-time.

We see, (1.5), that both the commutator and the anti-commutator are non-zero. These are not chiral algebras.

Division algebras derived from non-commutative groups do not interest us in this book.

Algebras from commutative groups:

Within a commutative group, the order of the elements multiplied together does not affect the product variable. For example, within any commutative group with three elements, $\{B, C, D\}$, we will have:

$$BC = D \qquad\qquad CB = D \qquad\qquad (1.6)$$

All non-commutative groups underlie only non-commutative division algebras, but it is only the great majority of commutative finite groups that underlie only commutative division algebras. Some commutative finite groups underlie non-commutative division algebras; there are very few such finite groups. We will be very much concerned with the commutative finite groups $C_2 \times C_2$ and $C_2 \times C_2 \times C_2$ both of which hold non-commutative division algebras[6].

Algebras with chirality:

There are two quaternion algebras which derive from the commutative finite group $C_2 \times C_2$. These two quaternion algebras are:

$$\mathbb{H}_{L\chi} = \begin{bmatrix} a & b & c & d \\ -b & a & -d & c \\ -c & d & a & -b \\ -d & -c & b & a \end{bmatrix} \qquad\qquad \mathbb{H}_{R\chi} = \begin{bmatrix} a & b & c & d \\ -b & a & d & -c \\ -c & -d & a & b \\ -d & c & -b & a \end{bmatrix} \qquad (1.7)$$

We refer to these algebras, (1.7), as the left-chiral quaternions and the right-chiral quaternions as denoted by the L or the R in the subscripts.

Because these two algebras derive from a commutative finite group, the two oppositely ordered products of any two of the separate variable matrices, $\{A, B, C, D\}$ is a matrix of the same variable. For example, the left-chiral quaternions have:

$$BC = +D \qquad\qquad CB = -D \qquad\qquad (1.8)$$

This is very different from the oppositely ordered products of an algebra derived from a non-commutative group. The non-commutativity of the quaternions is manifest in the sign of the variable that is in the product and not in the variable that is in the product. That's an important point to emphasize, and so we will repeat that point.

The non-commutativity of the quaternions is manifest in the sign of the product and not in the variable of the product. There are different types of non-commutativity.

---

[6] Note that the commutative group $C_2 \times C_6$ is known to hold non-commutative algebras.

Now have a cup of tea and let that fact soak into your brain; better still, go to the pub and give this fact a good soaking. It is important.

Chiral algebras in general:

With a little thought, we realise that the only way a commutative group can hold non-commutative algebras is if that non-commutativity is manifest in the sign of the product.

The anti-commutators and the commutator of the $B \& C$ variable matrices of the left-chiral quaternions are:

$$\{B \quad C\} = BC + CB = 0 \qquad\qquad [B \quad C] = BC - CB = 2D \qquad\qquad (1.9)$$

The anti-commutators and the commutator of the $B \& C$ variables matrices of the right-chiral quaternions are:

$$\{B \quad C\} = BC + CB = 0 \qquad\qquad [B \quad C] = BC - CB = -2D \qquad\qquad (1.10)$$

Within a quaternion algebra, there is one commutative element; this is the real variable on the leading diagonal of the algebraic matrix form. The anti-commutator and the commutator of this commutative variable with any of the imaginary quaternion variables are:

$$\{A \quad B\} = AB + BA = 2AB \qquad\qquad [A \quad B] = AB - BA = 0 \qquad\qquad (1.11)$$

We see that, within the quaternions, we always have either the anti-commutator or the commutator equal to zero for any choice of two variables.

Since the only way a commutative group can hold non-commutative algebras is if the non-commutativity of that algebra is manifest in the sign of the product, you might expect that every non-commutative algebra which derives from a commutative group is such that either the commutator of two variables is zero (the two variables commute) or the anti-commutator of two variables is zero (the two variables do not commute). Your expectation would be wrong. There are pairs of variables in the non-commutative 16-dimensional algebras which derive from the commutative $C_2 \times C_2 \times C_2 \times C_2$ group that have both non-zero commutators and non-zero anti-commutators.

Definition of chirality:

Algebras which are non-commutative in the way that the quaternions, $\mathbb{H}$, are non-commutative are chiral algebras. Commutative algebras like the complex numbers, $\mathbb{C}$, are not chiral algebras. Non-commutative algebras which are non-commutative in the way that the S_3 algebras are non-commutative are not chiral algebras.

Technically, chiral algebras are algebras in which all pairs of variables have anti-commutators and commutators of the forms (1.9) or (1.10) or (1.11).

It is important to realise that, for any given two variables in a chiral algebra, either the anti-commutator is zero or the commutator is zero. It is never the case within a chiral algebra that, for any given two variables, both the commutator and the anti-commutator are non-zero.

If, for two given variables within an algebra, both the commutator and the anti-commutator were non-zero, then the output of the anti-commutator would not be a 'whole variable'. We would have a pair of variables which are partially non-commutative and partially commutative. Technically, the commutator and the anti-commutator would not be defined, would not not exist, within such an algebra because the output of the commutator and the output of the anti-commutator would not be a variable within the algebra; there would be absence of closure under commutation and under anti-commutation.

What is chirality?

The conventional quaternion notation is:

$$\mathbb{H} = a + \hat{i}b + jc + kd \tag{1.12}$$

This notation hides the fact that there are both left-chiral quaternions and right-chiral quaternions.

The commutation relations of the left-chiral quaternions are:

$$\hat{i}j = k, \quad jk = \hat{i}, \quad k\hat{i} = j \quad : \quad j\hat{i} = -k, \quad kj = -\hat{i}, \quad \hat{i}k = -j \tag{1.13}$$

The commutation relations of the right-chiral quaternions are the opposite of these:

$$\hat{i}j = -k, \quad jk = -\hat{i}, \quad k\hat{i} = -j \quad : \quad j\hat{i} = k, \quad kj = \hat{i}, \quad \hat{i}k = j \tag{1.14}$$

In terms of the variables in (1.7), these relations are presented as commutators:

$$\begin{aligned} Left-chiral: \quad & [b \quad c] = 2d, \quad [c \quad d] = 2b, \quad [d \quad b] = 2c \\ Right-chiral: \quad & [b \quad c] = -2d, \quad [c \quad d] = -2b, \quad [d \quad b] = -2c \end{aligned} \tag{1.15}$$

Chirality is the direction of the commutation relations. The definition of the direction requires a reference frame; we will look at this later.

From where comes chirality?

To be a chiral algebra, a mathematical 'entity' needs to be a division algebra. All division algebras derive from finite groups. All division algebras derived from non-commutative finite groups have commutation relations like (1.5). Thus all division algebras derived from non-commutative finite groups are not chiral algebras. The division algebras derived from the commutative finite groups are almost always commutative division algebras, and the commutative division algebras are not chiral algebras.

We are left with the non-commutative division algebras which derive from the commutative finite groups; these are algebras like the quaternions whose non-commutativity is in the sign of the product. It is conjectured, but not yet proven, that the only commutative finite groups which hold non-commutative division algebras are the finite groups of the form $C_2 \times C_2$, $C_2 \times C_2 \times C_2$ and $C_2 \times C_2 \times C_2 \times ....$[7] These groups are of orders $2^n : n = 2, 3, 4, ...$

However, as we will see, only the order four group $C_2 \times C_2$ and the order eight group $C_2 \times C_2 \times C_2$ hold chiral algebras. We will see that the chirality 'breaks down' in the division algebras which derive from the order sixteen group $C_2 \times C_2 \times C_2 \times C_2$. Since the order sixteen $C_2 \times C_2 \times C_2 \times C_2$ group is a subgroup of all higher order $C_2 \times C_2 \times C_2 \times C_2 \times ...$ groups, chirality breaks down for these higher order groups.

The case of the $C_2 \times C_6$ group:

The non-commutativity of non-commutative algebras which derive from a commutative finite group is manifest in the sign of the product because, being derived from a commutative group, the variable in the product is the same regardless of the order of multiplication. This does not mean that all non-commutative algebras which derive from commutative groups are chiral algebras. The group $C_2 \times C_2 \times C_2 \times C_2$ is such a group in which the non-commutative algebras are not chiral.

The group $C_2 \times C_6$ is a commutative group which holds non-commutative algebras. Furthermore, the non-commutative algebras of the group $C_2 \times C_6$ are chiral algebras. However, the nature of the algebraic space of this group is such that it cannot be manifest in a space with only 2-dimensional rotations. Our 4-dimensional space-time has only 2-dimensional rotations. Thus it is that the chiral algebras of the $C_2 \times C_6$ group are not manifest in our 4-dimensional space-time.

Back to where we were:

In short, being a little tautological, chiral algebras which can be manifest in our 4-dimensional space-time exist in only the order four commutative group $C_2 \times C_2$ and the order eight commutative group $C_2 \times C_2 \times C_2$. Of course, this is quite astounding[8]. There are three algebraically distinct non-commutative division algebras within the $C_2 \times C_2 \times C_2$ group. There are two algebraically distinct non-commutative division algebras in the $C_2 \times C_2$ group. The algebras within these two groups are, except that there are five of them, completely unique. There are an infinity of division algebras within the infinity of finite groups, but these five of those algebras are 'extra special'. There are good reasons to think that these five algebras are manifest as our physical universe.

---

[7] And presumably any commutative groups which hold these groups as sub-groups; $C_2 \times C_6$ is such a commutative group which holds non-commutative algebras.

[8] There is no known mathematical proof that this statement is astounding, and so this statement is mere assertion.

Clifford algebras:

The division algebras of the $C_2 \times ...$ groups are very closely related to Clifford algebras. They are so closely related that, in this book, we will often refer to the division algebras of the $C_2 \times ...$ groups as the Clifford algebras[9], and we will use the terms interchangeably.

Why are we interested in chiral algebras?

The commutator and the anti-commutator play a fundamental role in quantum field theory and particle physics.

Lie algebras in particle physics:

The anti-commutators and commutators used in particle physics are within the Lie algebras $SU(2)$, $SU(3)$, and $SO(3,1)$. There is much that is unsatisfactory about Lie algebras in particle physics. The Lie group $SU(3)$ is so ugly that it cannot be the mathematics upon which the Great White Spirit chose to build the universe. There is no understanding of why only the $U(1)$, $SU(2)$, $SU(3)$, and $SO(3,1)$ Lie groups have a role in particle physics – why not also $SU(4)$ and $SU(5)$ and ...? Even $SU(2)$ is ugly compared with the quaternion rotation matrix.

The use of the Lie groups in gauge theory is also questionable. Within gauge theory, the Lie groups have a 'phase' which is really an angle; it is called a 'phase' because we are frightened to use the word angle within Lie algebra. Only division algebra spaces have angles[10].

The whole construction of Lie algebras and Lie groups is a mixture of arm waving and talk of infinitesimals that makes no sense. In short, Lie algebra and Lie groups are ugly. In her heart, the reader knows that the Great White Spirit would not use Lie algebras to construct the universe.

We therefore seek to understand the anti-commutator and the commutator using 'proper' mathematics. By 'proper' mathematics, we mean division algebras. Division algebras are beautiful and, being derived from no more than the existence of the number one, certainly exist[11]. We believe that the Great White Spirit used division algebras to build the universe. It is this belief that motivates this book.

Already, we have presented, but not explained, one result of interest; there is a limited number of chiral algebras which can be manifest in our 4-dimensional space-time. If we are asked, "Why did the Great White Spirit not use the $C_2 \times C_2 \times C_2 \times C_2$ algebras to build the universe?", we can answer, "Because they are not chiral algebras". If conventional particle physics is asked, "Why did the Great White Spirit not use the $SU(4)$ to build the universe?", it cannot answer.

Summary:

There are different types of non-commutativity.

---

[9] See: Dennis Morris : The Naked Spinor.
[10] The angles in our 4-dimensional space-time are angles from the 2-dimensional division algebras.
[11] See: Dennis Morris : Complex Numbers The Higher Dimensional Forms - 2nd Edition.

1) There is the non-commutativity we find in the non-commutative groups in which the order of multiplication determines the product variable.

2) There is the quaternion type of non-commutativity in which the product variable is unchanged by the order of multiplication but the multiplication is non-commutative because the sign of the product changes with the order of multiplication.

3) There is the mixed commutativity/non-commutativity of the 16-dimensional $C_2 \times C_2 \times C_2 \times C_2$ algebras. Who needs sci-fi when we have mathematics?

There are only five division algebras which have the quaternion type of non-commutativity and which can be manifest in our 4-dimensional space-time. There are chiral algebras which cannot be manifest in our 4-dimensional space-time.

The Commutation Reference Frame

In this chapter, we are going to pin down chirality. To do this, we need to produce a reference frame. To assist the reader, we reproduce the two quaternion algebras.

The quaternion algebras:
The two quaternion algebras are:

$$\mathbb{H}_{L\chi} = \begin{bmatrix} a & b & c & d \\ -b & a & -d & c \\ -c & d & a & -b \\ -d & -c & b & a \end{bmatrix} \qquad \mathbb{H}_{R\chi} = \begin{bmatrix} a & b & c & d \\ -b & a & d & -c \\ -c & -d & a & b \\ -d & c & -b & a \end{bmatrix} \qquad (2.1)$$

$$[b][c] = [d] \equiv \hat{i}\,j = k \qquad\qquad [b][c] = [-d] \equiv \hat{i}\,j = -k$$

We have included an example of the commutation relations. We have used the notation $[b]$ to indicate the matrix with the b variable etc.. This is the matrix remaining when the other variables in (2.1) are set to zero.

The chirality reference frame:
We are going to compare the commutation relations of chiral algebras. To be able to say that one algebra is right-chiral and the other algebra is left-chiral, we need a common algebraic reference frame. An algebraic reference frame is concerned with keeping the minus signs in the right places. This common reference frame will enable us to label one algebra left-chiral and another algebra right-chiral. For the 4-dimensional chiral algebras, we choose the common, but arbitrary, algebraic reference frame to be:

a) The variables in the algebraic matrix form are placed in the top row of the algebraic matrix form in alphabetic order from left to right.
b) There will be no minus signs on the top row of the algebraic matrix form.
c) We will consider the absolute values of the variables when deciding chirality.
d) The nature of multiplication of two variables is standard matrix multiplication from left to right.
e) The order of the variables being multiplied will be alphabetic from left to right.

We comment upon these conditions:

a) The first of these conditions is what we have done in every case above. It simply keeps our book-keeping tidy.
b) The absence of minus signs on the top row of the algebraic matrix form does not mean that the actual variables cannot be negative. We are concerned with the commutation relations; these will

naturally reverse for negative variables. We discuss this more when we consider conjugated algebraic matrix forms.

c) This is part of the second condition. By taking only the absolute values of the variables we avoid mixing chirality with the sign of the variable. After we have decided the chirality of the algebra, then we can let the values of the variables take any sign.

d) Standard left to right matrix multiplication is arbitrary but well established. For permutation matrices, the standard matrix multiplication is no more than sequential combination of permutations from right to left[12].

e) This arbitrary reference frame means that our anti-commutators and commutators will always be presented in alphabetic order from left to right. We will always have:

$$\{b \quad c\} \ \& \ [b \quad c] \tag{2.2}$$

We will never have:

$$\{c \quad b\} \ \& \ [c \quad b] \tag{2.3}$$

With these imposed conditions, we have an algebraic frame of reference and we can arbitrarily name one set of commutation relations to be left-chiral and thereby also name the set of opposite commutation relation to be right-chiral. Consider the left-chiral quaternion variables:

$$\begin{bmatrix} 0 & b & 0 & 0 \\ -b & 0 & 0 & 0 \\ 0 & 0 & 0 & -b \\ 0 & 0 & b & 0 \end{bmatrix} \begin{bmatrix} 0 & 0 & c & 0 \\ 0 & 0 & 0 & c \\ -c & 0 & 0 & 0 \\ 0 & -c & 0 & 0 \end{bmatrix} = \begin{bmatrix} 0 & 0 & 0 & bc \\ 0 & 0 & -bc & 0 \\ 0 & bc & 0 & 0 \\ -bc & 0 & 0 & 0 \end{bmatrix} \tag{2.4}$$

Looking at the form of the left-chiral quaternions above, (2.1), we take this multiplication, (2.4), to be left-chiral because it has produced a positive variable as the product; this can be seen by simply inspecting the sign of the variable on the top row of the product. This is equivalent to $\hat{i}\,j = k$ within the quaternions.

Consider the right-chiral quaternion variables:

$$\begin{bmatrix} 0 & b & 0 & 0 \\ -b & 0 & 0 & 0 \\ 0 & 0 & 0 & b \\ 0 & 0 & -b & 0 \end{bmatrix} \begin{bmatrix} 0 & 0 & c & 0 \\ 0 & 0 & 0 & -c \\ -c & 0 & 0 & 0 \\ 0 & c & 0 & 0 \end{bmatrix} = \begin{bmatrix} 0 & 0 & 0 & -bc \\ 0 & 0 & -bc & 0 \\ 0 & bc & 0 & 0 \\ bc & 0 & 0 & 0 \end{bmatrix} \tag{2.5}$$

Looking at the form of the right-chiral quaternions above, (2.1), we take this multiplication, (2.5), to be right-chiral because it has produced a negative variable in the product; this can be seen by simply inspecting the sign of the variable on the top row of the product. This is equivalent to $\hat{i}\,j = -k$ within the quaternions.

If we violate our reference frame by reversing the order of the left-chiral quaternion variables multiplication, (2.4), we get:

---

[12] See: Dennis Morris : Finite Groups A Simple Introduction.

$$\begin{bmatrix} 0 & 0 & c & 0 \\ 0 & 0 & 0 & c \\ -c & 0 & 0 & 0 \\ 0 & -c & 0 & 0 \end{bmatrix} \begin{bmatrix} 0 & b & 0 & 0 \\ -b & 0 & 0 & 0 \\ 0 & 0 & 0 & -b \\ 0 & 0 & b & 0 \end{bmatrix} = \begin{bmatrix} 0 & 0 & 0 & -bc \\ 0 & 0 & bc & 0 \\ 0 & -bc & 0 & 0 \\ bc & 0 & 0 & 0 \end{bmatrix} \tag{2.6}$$

This is equivalent to $\hat{ji} = -k$. The chirality is unchanged, but we have to wade through the confusion. To avoid this confusion, we have included the requirement that multiplication be in alphabetic order. We do not need this requirement, but it avoids confusion.

We are now able to say, arbitrarily, that the quaternion on the left of (2.1) is left-chiral and the quaternion on the right of (2.1) is of opposite chirality; it is right chiral.

Distribution of minus signs:
With careful thought, we realise that the sign of the top row variable in the ordered product is determined by the distribution of minus signs within the body of the matrix.

The chirality of an algebra is encompassed in the distribution of minus signs within the algebraic matrix form; that's worth repeating; the chirality of an algebra is encompassed in the distribution of minus signs within the algebraic matrix form of the algebra.

Summary so far:
The chirality of an algebra is the chirality of the commutation relations of that algebra. The chirality of an algebra is determined by the distribution of minus signs within the algebraic matrix form. This distribution is based upon all the elements of the top row of the algebraic matrix form being in alphabetic order and with positive sign.

Conjugates and chirality:
The chirality of an algebra is determined by the distribution of minus signs within the algebraic matrix form given the constraints of the reference frame listed above. A conjugate of a left-chiral quaternion is an element of the same left-chiral algebra as the left-chiral quaternion. The form of the distribution of the minus signs throughout the algebraic matrix form is the same, but the signs of the imaginary variables are reversed. We have the conjugate left-chiral quaternion:

$$\mathbb{H}_{L\chi}^{*} = \begin{bmatrix} a & -b & -c & -d \\ b & a & d & -c \\ c & -d & a & b \\ d & c & -b & a \end{bmatrix} \tag{2.7}$$

We have the product of two variables within the conjugate left-chiral quaternion:

$$\begin{bmatrix} 0 & -b & 0 & 0 \\ b & 0 & 0 & 0 \\ 0 & 0 & 0 & b \\ 0 & 0 & -b & 0 \end{bmatrix}\begin{bmatrix} 0 & 0 & -c & 0 \\ 0 & 0 & 0 & -c \\ c & 0 & 0 & 0 \\ 0 & c & 0 & 0 \end{bmatrix} = \begin{bmatrix} 0 & 0 & 0 & d \\ 0 & 0 & -d & 0 \\ 0 & d & 0 & 0 \\ -d & 0 & 0 & 0 \end{bmatrix} \tag{2.8}$$

We see that the product of two conjugate elements, multiplied together in the prescribed order, alphabetic, produces a positive sign on the top row. The chirality of the algebra is maintained under conjugation, but there is a subtlety here.

Two non-conjugate elements of a left-chiral quaternion multiplied together in the prescribed order produce a non-conjugate element as the product – see (2.4). Two conjugate elements of a left-chiral quaternion multiplied together in the prescribed order also produce a non-conjugate element as the product and do not produce a conjugate element as the product – see (2.8). In rather peculiar notation:

$$Conj_{L\chi} \times Conj_{L\chi} = Non-Conj_{L\chi}$$
$$Non-Conj_{L\chi} \times Non-Conj_{L\chi} = Non-Conj_{L\chi} \tag{2.9}$$

We have the conjugate right-chiral quaternion:

$$\mathbb{H}^*_{R\chi} = \begin{bmatrix} a & -b & -c & -d \\ b & a & -d & c \\ c & d & a & -b \\ d & -c & b & a \end{bmatrix} \tag{2.10}$$

The product of two conjugate right-chiral quaternion variables is:

$$\begin{bmatrix} 0 & -b & 0 & 0 \\ b & 0 & 0 & 0 \\ 0 & 0 & 0 & -b \\ 0 & 0 & b & 0 \end{bmatrix}\begin{bmatrix} 0 & 0 & -c & 0 \\ 0 & 0 & 0 & c \\ c & 0 & 0 & 0 \\ 0 & -c & 0 & 0 \end{bmatrix} = \begin{bmatrix} 0 & 0 & 0 & -d \\ 0 & 0 & -d & 0 \\ 0 & d & 0 & 0 \\ d & 0 & 0 & 0 \end{bmatrix} \tag{2.11}$$

We see that the product in the prescribed order of the two conjugate right-chiral elements of the right-chiral quaternion is a conjugate element rather than a non-conjugate element.

Looking at (2.5), we see that two non-conjugate elements of a right-chiral quaternion multiplied together also produce a conjugate element. In the same rather peculiar notation as we used above:

$$Conj_{R\chi} \times Conj_{R\chi} = Conj_{R\chi}$$
$$Non-Conj_{R\chi} \times Non-Conj_{R\chi} = Conj_{R\chi} \tag{2.12}$$

Notice the difference between (2.9) and (2.12).

We see that chirality is maintained by conjugation. This is just a small point, but it is nice to have tidied it.

Summary:

We have presented an algebraic reference frame which allows us to unambiguously assign a chirality to a non-commutative $C_2 \times C_2 \times ...$ algebra.

The chirality of an algebra is the chirality of the commutation relations of that algebra. The chirality of an algebra is determined by the distribution of minus signs within the algebraic matrix form. This distribution is based upon all the elements of the top row of the algebraic matrix form being in alphabetic order and with positive sign.

Conjugation preserves chirality.

Chapter 3

The 4-dimensional Chiral Algebras

We reiterate that the chiral algebras are non-commutative division algebras which derive from commutative finite groups[13]. The only known chiral algebras which can be manifest in our 4-dimensional space-time are those which derive from either the $C_2 \times C_2$ finite group or from the $C_2 \times C_2 \times C_2$ finite group. For the sake of completeness, we include a brief mention of the algebras of the C_2 finite group in this chapter.

The 2-dimensional algebras:

The general algebraic matrix form of the 2-dimensional division algebras is based upon the Standard Form Cayley table[14] of the C_2 finite group. We have:

$$\text{Cayley Table} \sim C_2 = \begin{bmatrix} a & b \\ b & a \end{bmatrix} \qquad \text{Algebraic matrix form} \sim C_2 = \begin{bmatrix} a & b \\ P_{2,1}b & a \end{bmatrix} \qquad (3.1)$$

The two 2-dimensional division algebras, the complex numbers, $\mathbb{C}$, and the hyperbolic complex numbers, $\mathbb{S}$, are formed by setting the parameter, $P_{2,1}$ in (3.1), equal to either minus one or plus one respectively and by then taking the matrix exponential. Note that the parameter, as with all such parameters, can be given any non-zero value; it is the sign of the parameter that separates the types of algebra. We have:

$$\mathbb{C} = \begin{bmatrix} a & b \\ -b & a \end{bmatrix} \equiv \exp\left(\begin{bmatrix} a & b \\ -b & a \end{bmatrix} \right) = \begin{bmatrix} r & 0 \\ 0 & r \end{bmatrix} \begin{bmatrix} \cos b & \sin b \\ -\sin b & \cos b \end{bmatrix} \qquad : \quad \{r = e^a\}$$

$$\mathbb{S} = \exp\left(\begin{bmatrix} a & b \\ b & a \end{bmatrix} \right) = \begin{bmatrix} h & 0 \\ 0 & h \end{bmatrix} \begin{bmatrix} \cosh b & \sinh b \\ \sinh b & \cosh b \end{bmatrix} \qquad : \quad \{h = e^a\}$$

$$(3.2)$$

We reiterate that the Euclidean complex numbers, $\mathbb{C}$, are a division algebra in both the Cartesian form and the polar form but that the hyperbolic complex numbers, $\mathbb{S}$, are a division algebra in only their polar form.

---

[13] This does not mean that all non-commutative division algebras which derive from commutative finite groups are chiral algebras.

[14] The Standard Form Cayley table is a Cayley table with the identity down the leading diagonal. There is most often more than one form of The Standard Form Cayley table for a particular group.

Into 4-dimensions:

The general matrix forms of the 4-dimensional division algebras is based upon the Standard Form Cayley tables of the two order four finite groups. We have no concern with the C_4 cyclic finite group, and so we give the Standard Form Cayley table of only the $C_2 \times C_2$ finite group. We have:

$$C_2 \times C_2 = \begin{bmatrix} a & b & c & d \\ b & a & d & c \\ c & d & a & b \\ d & c & b & a \end{bmatrix} \tag{3.3}$$

There are two algebraic matrix forms based upon the $C_2 \times C_2$ Standard Form Cayley table[15]. These are the commutative form and the non-commutative form:

$$\text{Commutative form} = \begin{bmatrix} a & b & c & d \\ P_{2,1}b & a & \dfrac{P_{2,1}}{P_{2,4}}d & P_{2,4}c \\ P_{3,1}c & \dfrac{P_{3,1}}{P_{2,4}}d & a & P_{2,4}b \\ \dfrac{P_{2,1}P_{3,1}}{P_{2,4}^2}d & \dfrac{P_{3,1}}{P_{2,4}}c & \dfrac{P_{2,1}}{P_{2,4}}b & a \end{bmatrix} \tag{3.4}$$

And:

$$\text{Non-Commutative form} = \begin{bmatrix} a & b & c & d \\ P_{2,1}b & a & \dfrac{P_{2,1}}{P_{2,4}}d & P_{2,4}c \\ P_{3,1}c & -\dfrac{P_{3,1}}{P_{2,4}}d & a & -P_{2,4}b \\ -\dfrac{P_{2,1}P_{3,1}}{P_{2,4}^2}d & \dfrac{P_{3,1}}{P_{2,4}}c & -\dfrac{P_{2,1}}{P_{2,4}}b & a \end{bmatrix} \tag{3.5}$$

The difference between (3.4) and (3.5) is four minus signs.

The commutative C_2 x C_2 algebras:

We briefly mention the commutative $C_2 \times C_2$ algebras. The 4-dimensional commutative algebraic matrix form, (3.4), holds eight commutative algebras of which two are the fully symmetric A_1 algebras and six are the A_2 algebras that each have one symmetric imaginary variable and two anti-symmetric imaginary variables.

---

[15] See : Dennis Morris : The Physics of Empty Space.

The two A_1 algebras are:

$$A_1^{\text{Case 1}} = \exp \left(\begin{bmatrix} a & b & c & d \\ b & a & d & c \\ c & d & a & b \\ d & c & b & a \end{bmatrix} \right) \qquad A_1^{\text{Case 2}} = \exp \left(\begin{bmatrix} a & b & c & d \\ b & a & -d & -c \\ c & -d & a & -b \\ d & -c & -b & a \end{bmatrix} \right) \tag{3.6}$$

$$P_{2,1} = P_{2,4} = P_{3,1} = +1 \qquad\qquad P_{2,1} = P_{3,1} = +1, \; P_{2,4} = -1$$

Briefly note that these two separate algebraic matrix forms, (3.6), do not commute with each other under matrix multiplication.

Since we associate chirality with commutation relations, the A_1 commutative algebras are not chiral algebras, but we briefly note that the variables of the A_1 algebras are related as:

$$\begin{array}{ll} \text{Case 1} & \text{Case 2} \\ bc = d & bc = -d \\ cb = d & cb = -d \end{array} \tag{3.7}$$

The A_2 algebraic matrix forms are:

$$\begin{bmatrix} a & b & c & d \\ b & a & d & c \\ -c & -d & a & b \\ -d & -c & b & a \end{bmatrix} \qquad\qquad \begin{bmatrix} a & b & c & d \\ b & a & -d & -c \\ -c & d & a & -b \\ -d & c & -b & a \end{bmatrix} \tag{3.8}$$

$$P_{2,1} = P_{2,4} = +1, \; P_{3,1} = -1 \qquad P_{2,4} = P_{3,1} = -1, \; P_{2,1} = +1$$

$$\begin{bmatrix} a & b & c & d \\ -b & a & -d & c \\ -c & -d & a & b \\ d & -c & -b & a \end{bmatrix} \qquad\qquad \begin{bmatrix} a & b & c & d \\ -b & a & d & -c \\ -c & d & a & -b \\ d & c & b & a \end{bmatrix} \tag{3.9}$$

$$P_{2,1} = P_{3,1} = -1, \; P_{2,4} = +1 \qquad P_{2,4} = P_{3,1} = P_{2,1} = -1$$

$$\begin{bmatrix} a & b & c & d \\ -b & a & d & -c \\ c & -d & a & -b \\ -d & -c & b & a \end{bmatrix} \qquad\qquad \begin{bmatrix} a & b & c & d \\ -b & a & -d & c \\ c & d & a & b \\ -d & c & -b & a \end{bmatrix} \tag{3.10}$$

$$P_{2,1} = P_{2,4} = -1, \; P_{3,1} = +1 \qquad P_{2,4} = P_{3,1} = +1, \; P_{2,1} = -1$$

The A_2 algebraic matrix forms come in pairs with similar relations to the A_1 algebraic matrix forms as expressed in (3.7). Note that we need to take the exponential of the algebraic matrix forms to form the division algebras.

These commutative algebras are of little interest to us.

The non-commutative C_2 x C_2 algebras:

The 4-dimensional non-commutative algebraic matrix form, (3.5), holds eight algebras of which two are the fully anti-symmetric quaternion algebras and six are the A_3 algebras that each have one anti-symmetric imaginary variable and two symmetric imaginary variables.

We have the two quaternion algebras:

$$\mathbb{H}_{L\chi} = \begin{bmatrix} a & b & c & d \\ -b & a & -d & c \\ -c & d & a & -b \\ -d & -c & b & a \end{bmatrix} \qquad \mathbb{H}_{R\chi} = \begin{bmatrix} a & b & c & d \\ -b & a & d & -c \\ -c & -d & a & b \\ -d & c & -b & a \end{bmatrix} \qquad (3.11)$$

We also have the six A_3 algebras:

$$SSA_{L\chi} = \exp\left(\begin{bmatrix} a & b & c & d \\ b & a & -d & -c \\ c & d & a & b \\ -d & -c & b & a \end{bmatrix}\right) \qquad SSA_{R\chi} = \exp\left(\begin{bmatrix} a & b & c & d \\ b & a & d & c \\ c & -d & a & -b \\ -d & c & -b & a \end{bmatrix}\right) \qquad (3.12)$$

$$SAS_{L\chi} = \exp\left(\begin{bmatrix} a & b & c & d \\ b & a & d & c \\ -c & d & a & -b \\ d & -c & -b & a \end{bmatrix}\right) \qquad SAS_{R\chi} = \exp\left(\begin{bmatrix} a & b & c & d \\ b & a & -d & -c \\ -c & -d & a & b \\ d & c & b & a \end{bmatrix}\right) \qquad (3.13)$$

$$ASS_{L\chi} = \exp\left(\begin{bmatrix} a & b & c & d \\ -b & a & -d & c \\ c & -d & a & -b \\ d & c & b & a \end{bmatrix}\right) \qquad ASS_{R\chi} = \exp\left(\begin{bmatrix} a & b & c & d \\ -b & a & d & -c \\ c & d & a & b \\ d & -c & -b & a \end{bmatrix}\right) \qquad (3.14)$$

The chirality of the quaternions:

We note that we have presented the quaternions, (3.11), and the A_3 algebras, (3.12) & (3.13) & (3.14), in accordance with the algebraic reference frame outlined in a previous chapter. We have presented the chirality of the quaternions previously, but we present it again for the convenience of the reader.

The commutation relations of the left-chiral quaternions are:

$$\hat{i}\hat{j} = \hat{k}, \qquad \hat{j}\hat{k} = \hat{i}, \qquad \hat{k}\hat{i} = \hat{j} \qquad : \qquad \hat{j}\hat{i} = -\hat{k}, \qquad \hat{k}\hat{j} = -\hat{i}, \qquad \hat{i}\hat{k} = -\hat{j} \qquad (3.15)$$

The commutation relations of the right-chiral quaternions are the opposite of these:

$$\hat{i}\hat{j} = -\hat{k}, \qquad \hat{j}\hat{k} = -\hat{i}, \qquad \hat{k}\hat{i} = -\hat{j} \qquad : \qquad \hat{j}\hat{i} = \hat{k}, \qquad \hat{k}\hat{j} = \hat{i}, \qquad \hat{i}\hat{k} = \hat{j} \qquad (3.16)$$

In terms of the variables in (3.11), these relations are presented as commutators:

$$Left - chiral: \qquad \begin{bmatrix} b & c \end{bmatrix}^{\mathbb{H}}_{L\chi} = 2d, \qquad \begin{bmatrix} c & d \end{bmatrix}^{\mathbb{H}}_{L\chi} = 2b, \qquad \begin{bmatrix} d & b \end{bmatrix}^{\mathbb{H}}_{L\chi} = 2c$$

$$Right - chiral: \qquad \begin{bmatrix} b & c \end{bmatrix}^{\mathbb{H}}_{R\chi} = -2d, \qquad \begin{bmatrix} c & d \end{bmatrix}^{\mathbb{H}}_{R\chi} = -2b, \qquad \begin{bmatrix} d & b \end{bmatrix}^{\mathbb{H}}_{R\chi} = -2c$$

$$(3.17)$$

Chirality is the direction of the commutation relations.

The chirality of the A_3 SSA algebras:

Although the exponential is central to the formulation of this algebraic matrix form as a division algebra, for the purpose of chirality, we ignore the need to take the exponential.

We have arbitrarily designated the above $SSA_{R\chi}$ A_3 algebra to be right-chiral and the $SSA_{L\chi}$ algebra to be left-chiral. The reader might think that we should have endeavoured to assign the arbitrary chirality to these algebras in a way that matched the assigned chirality of the quaternions; this cannot be done. Each type of algebra has its own type of chirality. This is an important point, and so we reiterate it.

Each type of algebra has its own type of chirality.

Every division algebra is associated with a division algebra space similarly to the way the 2-dimensional complex numbers, $\mathbb{C}$, are associated with the 2-dimensional complex plane, $\mathbb{C}$. Each type of division algebra space has its own type of rotation expressed as its own type of rotation matrix with its own type of trigonometric functions. Similarly, each type of division algebra space has its own type of chirality.

In practice, it is the sign of the commutator which interests us; we therefore often omit the 2. The full set of commutation relations of the right-chiral $SSA_{R\chi}$ algebra, are:

$$\begin{bmatrix} b & c \end{bmatrix}^{SSA}_{R\chi} = +d \qquad \begin{bmatrix} c & d \end{bmatrix}^{SSA}_{R\chi} = -b \qquad \begin{bmatrix} b & d \end{bmatrix}^{SSA}_{R\chi} = +c$$

$$\begin{bmatrix} c & b \end{bmatrix}^{SSA}_{R\chi} = -d \qquad \begin{bmatrix} d & c \end{bmatrix}^{SSA}_{R\chi} = +b \qquad \begin{bmatrix} d & b \end{bmatrix}^{SSA}_{R\chi} = -c$$

$$(3.18)$$

We have naughtily included the brackets with reverse alphabetic order variables just to ... well, we don't know why we did it[16].

If we associate the variables of the right-chiral $SSA_{R\chi}$ algebra, (3.12), with the imaginary elements:

$$\hat{i} = b \qquad \hat{j} = c \qquad \hat{k} = d \qquad (3.19)$$

---

[16] We did it because much experience of studying difficult and obscure mathematics has inclined your author to be very sympathetic to the studious reader's difficulty in comprehending this complicated area of mathematics. Technically, we should not have been naughty, but if it clarifies things for the reader, then ...

We have the right-chiral commutation relations:

$$SSA_{R\chi} \qquad \begin{array}{lll} \hat{i}j = k & jk = -\hat{i} & \hat{i}k = j \\ j\hat{i} = -k & kj = \hat{i} & k\hat{i} = -j \end{array} \qquad (3.20)$$

The reader might like to compare this, (3.20), to the quaternion case, (3.15) & (3.16).

The commutation relations of the left-chiral $SSA_{L\chi}$ algebra, (3.12), are:

$$[b \quad c]^{SSA}_{L\chi} = -d \qquad [c \quad d]^{SSA}_{L\chi} = +b \qquad [b \quad d]^{SSA}_{L\chi} = -c \qquad (3.21)$$

We see that the commutation relations of the left-chiral algebra are the reverse of the commutation relations of the right-chiral algebra.

The other A₃ algebras:

The other A_3 algebras have the commutation relations:

$$[b \quad c]^{SAS}_{L\chi} = +d \qquad [c \quad d]^{SAS}_{L\chi} = +b \qquad [b \quad d]^{SAS}_{L\chi} = +c \qquad (3.22)$$

$$[b \quad c]^{SAS}_{R\chi} = -d \qquad [c \quad d]^{SAS}_{R\chi} = -b \qquad [b \quad d]^{SAS}_{R\chi} = -c \qquad (3.23)$$

And:

$$[b \quad c]^{ASS}_{L\chi} = +d \qquad [c \quad d]^{ASS}_{L\chi} = -b \qquad [b \quad d]^{ASS}_{L\chi} = -c \qquad (3.24)$$

$$[b \quad c]^{ASS}_{R\chi} = -d \qquad [c \quad d]^{ASS}_{R\chi} = +b \qquad [b \quad d]^{ASS}_{R\chi} = +c \qquad (3.25)$$

Okay, in the *SAS* case, we did choose the arbitrary left-chirality of the A_3 algebras to match the quaternions after all.

Summary:

Each type of chiral division algebra has its own set of commutation relations, and so each chiral division algebra has its own type of chirality.

There are two algebraically non-isomorphic types of non-commutative 4-dimensional division algebra; these are the quaternions and the A_3 algebras. There are two quaternion algebras and six A_3 algebras. There is a left-chiral quaternion algebra, and there is a right-chiral quaternion algebra. The six A_3 algebras are in three pairs; each pair of A_3 algebras is a left-chiral algebra and a right-chiral algebra.

Tables:

In these tables, we use the abbreviation bc to indicate the commutator $[b \quad c] = bc - cb$. We do this because, when we get to 8-dimensions, we need the room in the tables

Commutation Relations of the quaternion algebras			
	bc	bd	cd
$\mathbb{H}_{L\chi}$	$+d$	$-c$	$+b$
$\mathbb{H}_{R\chi}$	$-d$	$+c$	$-b$
Commutation Relations of the A_3 algebras			
	bc	bd	cd
$SSA_{L\chi}$	$-d$	$-c$	$+b$
$SSA_{R\chi}$	$+d$	$+c$	$-b$
$SAS_{L\chi}$	$+d$	$+c$	$+b$
$SAS_{R\chi}$	$-d$	$-c$	$-b$
$ASS_{L\chi}$	$+d$	$-c$	$-b$
$ASS_{R\chi}$	$-d$	$+c$	$+b$

Chapter 4

Lie Group Commutation Relations

This chapter is included to tidy a few odd ends. The reader can skip this chapter if they choose to do so without interrupting the remainder of the book.

Commutation relations and Lie groups:
The commutation relations of the form:

$$[b \quad c] = -[c \quad b] \tag{4.1}$$

are most familiar to particle physicists through the Lie groups. Lie groups were named after the Norwegian mathematician Sophus Lie (1842-1899). The Lie groups $U(1)$, $SU(2)$ & $SU(3)$ are used to form the Standard Model of particle physics. The Lie group $SO(3,1)$, also called the Lorentz group, plays no part in the Standard Model but is taken to be the commutation relations of our 4-dimensional space-time. The Lie group $SO(3)$ also plays no part in the Standard Model but is taken to be the commutation relations of the spatial 3-dimensional part of our 4-dimensional space-time.

The Lie group U(1):
The Lie group $U(1)$[17] is just the unit circle in the 2-dimensional complex plane, $\mathbb{C}$:

$$U(1): \qquad e^{i\theta} \equiv \begin{bmatrix} \cos\theta & \sin\theta \\ -\sin\theta & \cos\theta \end{bmatrix} \tag{4.2}$$

The complex numbers, $\mathbb{C}$, is a commutative division algebra; as such, it has no commutation relations. Within gauge theory, the electromagnetic interaction[18] corresponds to a locally varying[19] phase within the $U(1)$ Lie group. Casting aside the smoke and mirrors, a phase within the $U(1)$ Lie group is just an angle in the complex plane.

The Lie group SU(2):
The Lie group $SU(2)$[20] is the gauge group of the weak interaction[21].

The standard mantra is, "The commutation relations of the Lie group $SU(2)$ are isomorphic as a Lie group to the commutation relations of the quaternions". Not quite; there is no chirality in Lie groups.

---

[17] This is the Unitary 1 group. Unitary means it has determinant of unity.
[18] The electromagnetic interaction is the same thing as the electromagnetic force.
[19] Locally varying means it varies from point to point in our 4-dimensional space-time.
[20] This is the Special Unitary 2 group. Special Unitary means it has determinant of plus unity.
[21] The weak interaction is the weak nuclear force.

None-the-less, the two sets of commutation relations are identical if we have no concept of chirality as is the case in Lie algebra.

We see there are two ways in mathematics to arrive at the $SU(2)$/quaternion commutation relations. We find these commutation relations in the quaternions, and we find these commutation relations in the Lie group $SU(2)$. Let me write with candour; it is my opinion that particle physics should dump Lie algebra and prefer instead the division algebras of the $C_2 \times C_2$ finite group and the $C_2 \times C_2 \times C_2$ finite group. In particular, I opine that the Lie group $U(1)$ should be replaced by the complex numbers, $\mathbb{C}$, and that the Lie group $SU(2)$ should be replaced by the quaternions. Having expressed my opinion, let us proceed.

We have seen elsewhere[22] that rotations exist in multi-dimensional forms and not in only 2-dimensional forms. We present the left-chiral quaternion rotation matrix as an example of a 4-dimensional rotation:

$$\exp\left(\begin{bmatrix} a & b & c & d \\ -b & a & -d & c \\ -c & d & a & -b \\ -d & -c & b & a \end{bmatrix}\right) = \begin{bmatrix} e^a & 0 & 0 & 0 \\ 0 & e^a & 0 & 0 \\ 0 & 0 & e^a & 0 \\ 0 & 0 & 0 & e^a \end{bmatrix} \begin{bmatrix} \cos\lambda & \dfrac{b}{\lambda}\sin\lambda & \dfrac{c}{\lambda}\sin\lambda & \dfrac{d}{\lambda}\sin\lambda \\ -\dfrac{b}{\lambda}\sin\lambda & \cos\lambda & -\dfrac{d}{\lambda}\sin\lambda & \dfrac{c}{\lambda}\sin\lambda \\ -\dfrac{c}{\lambda}\sin\lambda & \dfrac{d}{\lambda}\sin\lambda & \cos\lambda & -\dfrac{b}{\lambda}\sin\lambda \\ -\dfrac{d}{\lambda}\sin\lambda & -\dfrac{c}{\lambda}\sin\lambda & \dfrac{b}{\lambda}\sin\lambda & \cos\lambda \end{bmatrix} \quad (4.3)$$

$$\lambda = \sqrt{b^2 + c^2 + d^2}$$

The functions within the rotation matrix are the quaternion trigonometric functions. The argument of the trigonometric functions is the 4-dimensional quaternion angle of quaternion space. We see that the 4-dimensional angle takes three variables. The familiar 2-dimensional trigonometric functions take an angle of one variable; an example is (4.2); the 3-dimensional trigonometric functions[23] take an angle of two variables etc..

Our physical experience accustoms us to only 2-dimensional angles. Our 4-dimensional space-time is not a division algebra space. A 4-dimensional division algebra space, like quaternion space, has one real axis and three imaginary axes and holds only one 4-dimensional angle and has only one 4-dimensional rotation matrix. Our 4-dimensional space-time holds six 2-dimensional angles in six 2-dimensional rotation matrices. When Sophus Lie formulated Lie algebras, the higher dimensional types of angle were not known and it was presumed that there are only 2-dimensional angles. Thus, Lie groups are sets of 2-dimensional rotations. The Lie group $SU(2)$ is the set of three 2×2 matrices with determinant equal to plus one. These three matrices each generate a rotation matrix when the exponential of the matrix is taken. This is three 2-dimensional rotations.

The difference between the Lie group $SU(2)$ and the quaternion rotation matrix is not the commutation relations which are the same for both $SU(2)$ and the quaternion rotation matrix. The difference between $SU(2)$ and the quaternion rotation matrix is that $SU(2)$ is a set of three 2-dimensional rotations whereas the quaternion rotation matrix is a single 4-dimensional rotation. The quaternions are also a division

---

[22] See: Dennis Morris : Complex Numbers The Higher Dimensional Forms – 2nd Edition.
[23] See: Dennis Morris : Complex Numbers The Higher Dimensional Forms – 2nd Edition.

algebra whereas the Lie algebras are not 'really' algebras but are constructed from ill-defined infinitesimals and a lot of arm waving.

If we use $SU(2)$ as a gauge space over our 4-dimensional space-time, the locally varying phase is three locally varying 2-dimensional angles. If we use the quaternions as a gauge space over our 4-dimensional space-time, the locally varying phase is a single locally varying 4-dimensional angle in quaternion space.

The quaternion space has a rotation matrix, distance function, angles, an E-field and a B-field and 'Maxwell type equations'. The Lie group $SU(2)$ has none of these attributes. The reader will form their own opinion regarding whether or not the Lie group $SU(2)$ should be replaced by the quaternion rotation matrix.

The Lie group SU(3):

The Lie group $SU(3)$[24] is the gauge group of the strong nuclear force.

Given the above, it might seem sensible to seek to replace the Lie group $SU(3)$ with an algebra from the $C_2 \times C_2 \times C_2$ finite group. There is no algebra from the $C_2 \times C_2 \times C_2$ finite group which has commutation relations isomorphic as a Lie group to $SU(3)$. Indeed, $SU(3)$ has eight generator matrices generating eight 2-dimensional rotations, but there are no $C_2 \times C_2 \times C_2$ algebras which have commutation relations generated by eight distinct non-commutative elements.

In gauge theory, the gauge groups are seen as 'spaces' in which, at the very least, we have a phase – a 2-dimensional angle. If we think of a Lie group as a space in which all rotations are 2-dimensional, then we will have as many rotational planes as there are pairs of axes. 2-dimensional space has one pair of axes; 3-dimensional space has three pairs of axes; 4-dimensional space has six pairs of axes; 5-dimensional space has ten pairs of axes. The single generator matrix of $U(1)$ corresponds to the single pair of axes in 2-dimensional space; the three generator matrices of $SU(2)$ correspond to the three pairs of axes in 3-dimensional space. There is no space of any dimension that has eight pairs of axes. Since $SU(3)$ has eight generator matrices, it cannot be a space of 2-dimensional rotations in the way that the Lie groups $U(1)$ and $SU(2)$ can each be a space of 2-dimensional rotations.

The Standard Model of particle physics presumes there are eight gluons because $SU(3)$ has eight generator matrices and therefore has eight 'phases' to vary locally. There is no observational evidence of the number of gluons, and so we can proceed on the basis that there might be more than eight gluons or less than eight gluons.

The commutation relations of the Lie group SU(2):

In the standard basis, the Lie group $SU(2)$ is represented by the three matrices:

---

[24] This is the Special Unitary 3 group. Special Unitary means it has determinant of plus unity.

$$J_1 = \begin{bmatrix} 0 & 1 \\ 1 & 0 \end{bmatrix}, \qquad J_2 = \begin{bmatrix} 0 & -i \\ i & 0 \end{bmatrix}, \qquad J_3 = \begin{bmatrix} 1 & 0 \\ 0 & -1 \end{bmatrix} \tag{4.4}$$

We have the commutation relations:

$$\begin{aligned}
[J_1 \quad J_2] &= \begin{bmatrix} 0 & 1 \\ 1 & 0 \end{bmatrix}\begin{bmatrix} 0 & -i \\ i & 0 \end{bmatrix} - \begin{bmatrix} 0 & -i \\ i & 0 \end{bmatrix}\begin{bmatrix} 0 & 1 \\ 1 & 0 \end{bmatrix} \\
&= \begin{bmatrix} i & 0 \\ 0 & -i \end{bmatrix} - \begin{bmatrix} -i & 0 \\ 0 & i \end{bmatrix} \\
&= 2\begin{bmatrix} i & 0 \\ 0 & -i \end{bmatrix} = 2iJ_3
\end{aligned} \tag{4.5}$$

and:

$$[J_1 \quad J_3] = -2iJ_2 \qquad\qquad [J_2 \quad J_3] = 2iJ_1 \tag{4.6}$$

The reader will see that, if we ignore the unsightly i, these commutation relations match the commutation relations of the left-chiral quaternions.

Within a Lie algebra, there is the concept of a Casimir operator which commutes with all the elements of the Lie algebra. Within $SU(2)$, there is one Casimir operator which is:

$$J^2 = J_1^2 + J_2^2 + J_3^2 = \begin{bmatrix} 1 & 0 \\ 0 & 1 \end{bmatrix} + \begin{bmatrix} 1 & 0 \\ 0 & 1 \end{bmatrix} + \begin{bmatrix} 1 & 0 \\ 0 & 1 \end{bmatrix} = 3\begin{bmatrix} 1 & 0 \\ 0 & 1 \end{bmatrix} \tag{4.7}$$

This is, of course, just a real number. Comparing this to the quaternions, we have:

$$\hat{i}^2 + \hat{j}^2 + \hat{k}^2 = -3 \tag{4.8}$$

The commutation relations of the Lie group SU(3):

In the standard basis, the Lie group $SU(3)$ is represented by the eight matrices known as the Gell-Mann matrices; these are:

$$\lambda_1 = \begin{bmatrix} 0 & 1 & 0 \\ 1 & 0 & 0 \\ 0 & 0 & 0 \end{bmatrix} \qquad\qquad \lambda_2 = \begin{bmatrix} 0 & -i & 0 \\ i & 0 & 0 \\ 0 & 0 & 0 \end{bmatrix} \tag{4.9}$$

$$\lambda_3 = \begin{bmatrix} 1 & 0 & 0 \\ 0 & -1 & 0 \\ 0 & 0 & 0 \end{bmatrix} \qquad\qquad \lambda_4 = \begin{bmatrix} 0 & 0 & 1 \\ 0 & 0 & 0 \\ 1 & 0 & 0 \end{bmatrix} \tag{4.10}$$

$$\lambda_5 = \begin{bmatrix} 0 & 0 & -i \\ 0 & 0 & 0 \\ i & 0 & 0 \end{bmatrix} \qquad \lambda_6 = \begin{bmatrix} 0 & 0 & 0 \\ 0 & 0 & 1 \\ 0 & 1 & 0 \end{bmatrix} \tag{4.11}$$

$$\lambda_7 = \begin{bmatrix} 0 & 0 & 0 \\ 0 & 0 & -i \\ 0 & i & 0 \end{bmatrix} \qquad \lambda_8 = \frac{1}{\sqrt{3}} \begin{bmatrix} 1 & 0 & 0 \\ 0 & 1 & 0 \\ 0 & 0 & -2 \end{bmatrix} \tag{4.12}$$

We can see that nature would not choose these matrices because they are ugly; they have imaginary elements, a two, and a square root. You would have to be insane to think the Great White Spirit who watches over us all would have anything to do with this set of matrices. I certainly would not want to be watched over by any spirit, Great White or other, so without taste as to choose such an abomination as the eight matrices above (4.9) to (4.12) with which to build the universe. None-the-less, the Lie group $SU(3)$, as represented by the above eight matrices, (4.9) to (4.12), is a central part of the Standard Model of particle physics.

The $SU(3)$ commutation relations are calculated as:

$$\begin{aligned}
[\lambda_1 \quad \lambda_2] &= \lambda_1\lambda_2 - \lambda_2\lambda_1 \\
&= \begin{bmatrix} 0 & 1 & 0 \\ 1 & 0 & 0 \\ 0 & 0 & 0 \end{bmatrix}\begin{bmatrix} 0 & -i & 0 \\ i & 0 & 0 \\ 0 & 0 & 0 \end{bmatrix} - \begin{bmatrix} 0 & -i & 0 \\ i & 0 & 0 \\ 0 & 0 & 0 \end{bmatrix}\begin{bmatrix} 0 & 1 & 0 \\ 1 & 0 & 0 \\ 0 & 0 & 0 \end{bmatrix} \\
&= \begin{bmatrix} 2i & 0 & 0 \\ 0 & -2i & 0 \\ 0 & 0 & 0 \end{bmatrix} = 2i\lambda_3
\end{aligned} \tag{4.13}$$

We write this relation, (4.13), as:

$$[\lambda_1 \quad \lambda_2] = 2i\lambda_3 \tag{4.14}$$

Note that the commutator of the two elements of $SU(3)$ is another element of $SU(3)$ multiplied by a (complex) number.

We have:

$$[\lambda_1 \quad \lambda_2] = 2i\lambda_3 \qquad [\lambda_1 \quad \lambda_3] = -2i\lambda_2 \qquad [\lambda_2 \quad \lambda_3] = 2i\lambda_1 \tag{4.15}$$

We see that, ignoring the unsightly $2i$, the elements $\{\lambda_1, \lambda_2, \lambda_3\}$ are a $SU(2)$ sub-group of $SU(3)$[25].

We have:

---

[25] There are two other $SU(2)$ sub-groups in $SU(3)$.

$$[\lambda_1 \quad \lambda_4] = i\lambda_7 \qquad [\lambda_1 \quad \lambda_5] = -i\lambda_6 \qquad [\lambda_1 \quad \lambda_6] = i\lambda_5$$
$$[\lambda_1 \quad \lambda_7] = -i\lambda_4 \qquad [\lambda_1 \quad \lambda_8] = 0 \tag{4.16}$$

We see that λ_1 & λ_8 commute with each other.

We have:

$$[\lambda_2 \quad \lambda_4] = i\lambda_6 \qquad [\lambda_2 \quad \lambda_5] = i\lambda_7 \qquad [\lambda_2 \quad \lambda_6] = -i\lambda_4$$
$$[\lambda_2 \quad \lambda_7] = -i\lambda_5 \qquad [\lambda_2 \quad \lambda_8] = 0 \tag{4.17}$$

And:

$$[\lambda_3 \quad \lambda_4] = i\lambda_5 \qquad [\lambda_3 \quad \lambda_5] = -i\lambda_4 \qquad [\lambda_3 \quad \lambda_6] = -i\lambda_7$$
$$[\lambda_3 \quad \lambda_7] = i\lambda_6 \qquad [\lambda_3 \quad \lambda_8] = 0 \tag{4.18}$$

Until now, the commutators have been a single elements of $SU(3)$ multiplied by a number. However, we have:

$$[\lambda_4 \quad \lambda_5] = i(\lambda_3 + \lambda_8) \qquad [\lambda_4 \quad \lambda_6] = i\lambda_2$$
$$[\lambda_4 \quad \lambda_7] = i\lambda_1 \qquad [\lambda_4 \quad \lambda_8] = i\sqrt{3}\lambda_5 \tag{4.19}$$

We also have:

$$[\lambda_5 \quad \lambda_6] = -i\lambda_1 \qquad [\lambda_5 \quad \lambda_7] = i\lambda_2 \qquad [\lambda_5 \quad \lambda_8] = i\sqrt{3}\lambda_4 \tag{4.20}$$

And:

$$[\lambda_6 \quad \lambda_7] = -i(\lambda_3 - \sqrt{3}\lambda_8) \qquad [\lambda_6 \quad \lambda_8] = -i\sqrt{3}\lambda_7 \qquad [\lambda_7 \quad \lambda_8] = i\sqrt{3}\lambda_6 \tag{4.21}$$

The set of three elements $\{\lambda_2, \lambda_4, \lambda_6\}$ are a closed set under commutation. We have no identity element.

The two elements λ_3 & λ_8 commute with each other, but they do not both commute with every other element of $SU(3)$. Because they commute with each other, the two elements λ_3 & λ_8 are taken to be the Cartan sub-algebra of $SU(3)$.

Intriguingly, $SU(3)$ has two Casimir operators; that is two combinations of the elements of $SU(3)$ which commute with all the elements of $SU(3)$. The non-commutative 8-dimensional $C_2 \times C_2 \times C_2$ algebras also have two elements which commute with all the other elements of the algebra.

The Lie group SO(3,1):
We have now covered the Lie groups used in the Standard Model of particle physics. We will briefly cover the other Lie group that is used in physics.

We have shown elsewhere[26] that the emergent space of the six A_3 algebras is our 4-dimensional space-time. We have shown elsewhere[27] that the commutation relations of the Lorentz group $SO(3,1)$[28] match the commutation relations of the three left-chiral A_3 algebras taken together. The commutation relations of the Lorentz group $SO(3,1)$ also match the commutation relations of the three right-chiral A_3 algebras taken together. Thus, within the six A_3 algebras, we have a right-chiral copy of $SO(3,1)$ and a left chiral copy of $SO(3,1)$. We opine that this is why we have left-handedness and right-handedness in our 4-dimensional space-time.

Briefly, the SO(3) Lie groups:

The $SO(3)$[29] Lie groups are just three 2-dimensional rotations set into 3×3 matrices. We have:

$$\begin{bmatrix} \cos\theta & \sin\theta & 0 \\ -\sin\theta & \cos\theta & 0 \\ 0 & 0 & 1 \end{bmatrix} \quad \begin{bmatrix} \cos\phi & 0 & \sin\phi \\ 0 & 1 & 0 \\ -\sin\phi & 0 & \cos\phi \end{bmatrix} \quad \begin{bmatrix} 1 & 0 & 0 \\ 0 & \cos\psi & \sin\psi \\ 0 & -\sin\psi & \cos\psi \end{bmatrix} \quad (4.22)$$

These are rotations about an axis as can be shown by taking the eigenvectors of the matrices; in each case, one of the eigenvectors is independent of the angle of rotation – it is an axis of rotation. Technically, the group $SO(3)$ is the set of three generator matrices:

$$\begin{bmatrix} 0 & 1 & 0 \\ -1 & 0 & 0 \\ 0 & 0 & 0 \end{bmatrix} \quad \begin{bmatrix} 0 & 0 & 1 \\ 0 & 0 & 0 \\ -1 & 0 & 0 \end{bmatrix} \quad \begin{bmatrix} 0 & 0 & 0 \\ 0 & 0 & 1 \\ 0 & -1 & 0 \end{bmatrix} \quad (4.23)$$

Taking the exponentials of these matrices, (4.23), will give the rotation matrices above, (4.22).

Summary:

You do not need to understand Lie groups and Lie algebra to understand the remainder of this book.

The type of non-commutation based upon the sign of the product is found within the algebras of the $C_2 \times C_2$ and $C_2 \times C_2 \times C_2$ groups and within the Lie groups.

---

[26] See: Dennis Morris: Upon General Relativity.
[27] See: Dennis Morris: The Physics of Empty Space
[28] This is the Special Orthogonal group of 4-dimensional space-time.
[29] This is the Special Orthogonal group.

Chapter 5

The Neutrino Controversy

The neutrino mass controversy:

Neutrinos have been observed with great accuracy on many occasions to travel at the speed of light. No one has ever seen a neutrino moving at less than the speed of light. This implies that neutrinos are massless. The high end spectra of beta decay experiments indicate that neutrinos are massless, and there is other experimental evidence of neutrinos being massless. In short, there is a great deal of evidence that neutrinos are massless. The Standard Model of particle physics is predicated upon neutrinos being massless.

Neutrinos come in three types, or generations, called the electron neutrino, the muon neutrino, and the tau neutrino.

There is a huge amount of experimental evidence that neutrinos oscillate between generations. As neutrinos travel from the sun to the Earth, they change their type, their generation. There is a non-zero probability that an electron neutrino leaving the sun will have changed into a muon neutrino or a tau neutrino by the time it hits the Earth. Similarly, there is a non-zero probability that muon neutrinos leaving the sun will change into tau neutrinos or electron neutrinos as they travel from the sun to the Earth. The theory states that neutrinos can change generation, oscillate between the generations, only if the square of the neutrino mass is non-zero.

The neutrino mass controversy is that:

   a) Neutrinos move at the speed of light and therefore are massless.
   b) The neutrino mass squared must be non-zero because neutrinos oscillate between generations.

How can this be?

If the neutrino field is a quaternion field with zero real part, and we associate the real part of a quaternion with mass, then we have a solution. Taking the square of the neutrino field by multiplying the quaternion neutrino field by its quaternion conjugate produces a quaternion with non-zero real part. The squared neutrino field is massive

The Majorana neutrino or Dirac neutrino controversy:

The Dirac equation[30] is normally presented as being a equation in complex numbers. In particular, each term except the mass term is multiplied by $i\gamma$ where $i = \sqrt{-1} \in \mathbb{C}$ and γ is a 4×4 gamma matrix[31]. The Dirac equation presumes to describe the neutrino when the mass term is zero. Part of understanding the Dirac equation is that the fields of anti-matter particles are the complex conjugates of the fields of matter particles.

---

[30] See : Dennis Morris : The Quaternion Dirac Equation.
[31] Although the gamma matrices are written as 4×4 matrices, they are four of the sixteen basis elements of a 16-dimensional Clifford algebra.

In 1937, Ettore Majorana (1906-?)[32] wrote the gamma matrices in a basis such that every element within them was imaginary. The terms $i\gamma$ thus became real, and with that the Dirac equation with a zero mass term became a real number equation with only real solutions. Real solutions have no conjugates, and so the Dirac equation with a zero mass term does not describe anti-matter unless we accept that some matrix bases are special – which is silly. It was thus postulated that a neutrino is its own anti-particle.

We therefore have two proposed kinds of neutrino. A Dirac neutrino is massive and comes in distinct particle and anti-particle forms. A Majorana neutrino is massless and comes in only one form which is both particle and anti-particle. If the neutrino is massless, then the only difference between the two types of neutrino is the basis chosen for the gamma matrices – this is not pretty mathematics.

Controversy lingers as to whether or not neutrinos are Dirac neutrinos or Majorana neutrinos[33].

The left-handed neutrino:

We have kept the biggest neutrino controversy until last. The neutrino is left-handed. By left-handed, convention means that the axis of the intrinsic spin of the neutrino aligns in the opposite direction to the direction of the velocity of the neutrino. The standard mantra is that intrinsic spin is spin about an axis.

If the neutrino moved at less than the speed of light, as electrons do, then neutrinos could not be solely left-handed. To a stationary observer, the neutrino might be moving from left to right with its axis of intrinsic spin pointing from right to left. If that stationary observer was to get on her bike and start to move faster than the neutrino in a left to right direction, then, as seen by that moving observer as she overtakes the neutrino, the neutrino would be moving from right to left. The moving observer would see the neutrino's velocity to be in the same direction as the axis of its intrinsic spin; she would see a right-handed neutrino. To be always left-handed, the neutrino needs to move at the speed of light so that it cannot be overtaken by any observer; to do this, the neutrino needs to be massless.

There is a great deal of experimental evidence to support the solely left-handed nature of the neutrino. There is not a shred of evidence to dispute the solely left-handed nature of the neutrino except that, if neutrinos are massive, they cannot be solely left-handed. The observation of neutrino oscillations mentioned above are often taken to indicate that the neutrino is massive.

Intrinsic spin of the neutrino:

We opine that intrinsic spin exists in a spinor space[34]; in particular, for the electron and the neutrino, we opine that intrinsic spin exists in quaternion space. We opine that the two possible types of intrinsic spin, spin up and spin down, are actually the two chiralities of the quaternions. The reader will form their own opinion; things will be much clearer after the next couple of chapters.

---

[32] In 1938, Majorana disappeared. No one knows what became of him, but there is a report of him being in Peru in the 1950's.
[33] See: Dennis Morris: 'The Quaternion Dirac Equation' for a solution of this problem.
[34] See the next chapter.

Chapter 6

What is a Spinor?

In this chapter, we give a sweeping overview of the nature of spinors[35].

Division algebras:

We may define a division algebra in one of three ways:

1) A division algebra is usually defined as a mathematical system which satisfies the thirteen division algebra axioms. Note that multiplicative commutativity is not one of the thirteen division algebra axioms[36].

2) In this book, we define a division algebra as being a mathematical system which satisfies the thirteen division algebra axioms except that we do not require negative real numbers on the real number axis nor the additive identity, zero, on the real axis.

3) Alternatively, instead of talking of axioms, we can define a division algebra as a mathematical system which has all the properties of the hyperbolic complex numbers, $\mathbb{S}$, or the Euclidean complex numbers, $\mathbb{C}$, except that we allow non-commutative multiplication. The hyperbolic complex numbers are 2-dimensional space-time, and the Euclidean complex numbers are 2-dimensional Euclidean space, and so this is a 'physicist's definition'. This 'physicist's definition' concurs with the second definition immediately above.

Spinors:

The word spinor is used, and often abused, to mean many things within mathematics and physics[37]. Within this book, we will use the word spinor to mean an element of a n-dimensional type of complex numbers. By a type of complex number, we mean a type of division algebra; this is the same thing as a type of number. We say a spinor is an element of a spinor algebra. Well known examples of spinor algebras are the Euclidean complex numbers, $\mathbb{C}$, and the quaternions, $\mathbb{H}$. Thus, a quaternion is a spinor, and a Euclidean complex number is a spinor.

We use the terms 'division algebra', 'spinor algebra', and 'type of complex number' interchangeably. Associated with each type of n-dimensional complex number, there is a type of space which has one real axis and $(n-1)$ imaginary axes. The 2-dimensional complex plane, $\mathbb{C}$, is such a space, and it is associated with the 2-dimensional complex numbers division algebra, $\mathbb{C}$. The 4-dimensional quaternion space, $\mathbb{H}$, is such a space, and it is associated with the 4-dimensional complex numbers division algebra $\mathbb{H}$. These spaces are not $\mathbb{R}^n$ spaces. Such spaces of one real axis and $(n-1)$ imaginary axes are referred

---

[35] See: Dennis Morris : The Naked Spinor.
[36] A multiplicatively commutative division algebra is called an algebraic field.
[37] See : Dennis Morris : The Naked Spinor

to as 'spinor spaces', 'division algebra spaces', or 'complex number spaces'; sometimes, the term 'algebraic space' is used.

So, in short, a spinor is just an element of a n-dimensional number algebra.

There are an infinite number of different types of complex numbers[38]. There are the 1-dimensional real numbers, $\mathbb{R}$, which are often not considered to be complex numbers; of course, the real numbers are a division algebra. There are two types of 2-dimensional complex numbers; there are the Euclidean complex numbers, $\mathbb{C}$, which are often called just the complex numbers, and there are the hyperbolic complex numbers, $\mathbb{S}$, which are the 2-dimensional space-time of special relativity. There are four types of 3-dimensional complex numbers[39], and there are twenty-four types of 4-dimensional complex numbers including two quaternion algebras, $\mathbb{H}$, six A_3 algebras, six A_2 algebras, two A_1 algebras, and eight C_4 algebras in two algebraically non-isomorphic types. There are sixteen types of 5-dimensional complex number – many are algebraically isomorphic. There are one hundred and twenty-five non-algebraically isomorphic 15-dimensional types of complex numbers[40].

Spinors are often normalised to be of unit length by physicists and mathematicians, and so, if we wish, we can think of a spinor as being a unit length element of a n-dimensional complex number. In such a mind, a spinor is just a point on the unit sphere within the complex number space (spinor space). As such, a spinor is an element of a rotation matrix; for examples:

$$\mathbb{C}_{unit} = \begin{bmatrix} \cos\theta & \sin\theta \\ -\sin\theta & \cos\theta \end{bmatrix} \quad or \quad \mathbb{H}_{unit} = \begin{bmatrix} \cos\lambda & \frac{b}{\lambda}\sin\lambda & \frac{c}{\lambda}\sin\lambda & \frac{d}{\lambda}\sin\lambda \\ -\frac{b}{\lambda}\sin\lambda & \cos\lambda & -\frac{d}{\lambda}\sin\lambda & \frac{c}{\lambda}\sin\lambda \\ -\frac{c}{\lambda}\sin\lambda & \frac{d}{\lambda}\sin\lambda & \cos\lambda & -\frac{b}{\lambda}\sin\lambda \\ -\frac{d}{\lambda}\sin\lambda & -\frac{c}{\lambda}\sin\lambda & \frac{b}{\lambda}\sin\lambda & \cos\lambda \end{bmatrix} \quad (6.1)$$

$$\lambda = \sqrt{b^2 + c^2 + d^2}$$

Looking at the two rotation matrices above, (6.1), we see these are very different kinds of rotation matrices. Different kinds of rotation matrices imply different kinds of rotation; different kinds of rotation imply different kinds of angle. There is twice as much rotation in the quaternion rotation matrix as there is in the complex number rotation matrix; to see this more clearly, set to zero two of the variables in the quaternion rotation matrix. The quaternion spinor is said to be a 'double cover' of the Euclidean $\mathbb{C}$ spinor.

---

[38] See : Dennis Morris : Complex Numbers The Higher Dimensional Forms – 2nd Edition.
[39] Two of the four are algebraically isomorphic.
[40] See : Dennis Morris : The Uniqueness of our Space-time.

Other notation:

But I thought a spinor was an ordered pair of complex numbers. Within physics, a Pauli spinor is often presented, for historical reasons, as an ordered pair of complex numbers[41]. This is just an obscure way of writing a quaternion[42]; we have:

$$\psi_{Pauli} = \begin{bmatrix} \psi_1 \\ \psi_2 \end{bmatrix} = \begin{bmatrix} a+ib \\ c+id \end{bmatrix} = \begin{bmatrix} a & b & c & d \\ -b & a & -d & c \\ -c & d & a & -b \\ -d & -c & b & a \end{bmatrix} \in \mathbb{H} \tag{6.2}$$

A Dirac spinor, also called a bi-spinor, is simply an ordered pair of quaternions:

$$\psi_{Dirac} = \begin{bmatrix} \psi^1_{Pauli} \\ \psi^2_{Pauli} \end{bmatrix} = \begin{bmatrix} a+ib \\ c+id \\ e+if \\ g+ih \end{bmatrix} = \begin{bmatrix} \mathbb{H}_1 \\ \mathbb{H}_2 \end{bmatrix} \tag{6.3}$$

An ordered pair of quaternions is not a division algebra, and so a Dirac spinor does not exist within algebra[43].

An example of a spinor space:

From where come the spinor spaces? The spinor spaces come from the finite groups. We take the view that a finite group is no more than a closed set of permutations[44]. We write this set of permutations as a set of permutation matrices[45]. For example, the finite group C_2 is the two permutations written as permutation matrices:

$$\begin{bmatrix} 1 & 0 \\ 0 & 1 \end{bmatrix} \quad \& \quad \begin{bmatrix} 0 & 1 \\ 1 & 0 \end{bmatrix} \tag{6.4}$$

We multiply each of these permutation matrices by a real variable and add them to produce the algebraic matrix form:

$$\begin{bmatrix} a & 0 \\ 0 & a \end{bmatrix} + \begin{bmatrix} 0 & b \\ b & 0 \end{bmatrix} = \begin{bmatrix} a & b \\ b & a \end{bmatrix} \tag{6.5}$$

Taking the matrix exponential gives us the 2-dimensional spinor space which, in this case, is the 2-dimensional hyperbolic complex numbers, $\mathbb{S}$; this is the 2-dimensional space-time of special relativity:

---

[41] See : Dennis Morris : Even Mathematicians and Physicists make Mistakes.

[42] This mathematical fact has been known since at least 1929; possibly it was known to Hamilton in 1845, but it has been ignored by both physicists and mathematicians.

[43] See : Dennis Morris : The Quaternion Dirac Equation.

[44] See : Dennis Morris : Finite Groups A Simple Introduction.

[45] Permutation matrices are square matrices with a single 1 in each row and a single 1 in each column. Permutation matrices correspond on a one-to-one basis with permutations.

$$\exp\left(\begin{bmatrix} a & b \\ b & a \end{bmatrix}\right) = \begin{bmatrix} e^a & 0 \\ 0 & e^a \end{bmatrix}\begin{bmatrix} \cosh b & \sinh b \\ \sinh b & \cosh b \end{bmatrix} = \begin{bmatrix} e^a & 0 \\ 0 & e^a \end{bmatrix}\begin{bmatrix} \gamma & v\gamma \\ v\gamma & \gamma \end{bmatrix} \in \mathbb{S} \qquad (6.6)$$

With a minus sign, we also have the spinor space which is the complex plane, $\mathbb{C}$:

$$\exp\left(\begin{bmatrix} a & b \\ -b & a \end{bmatrix}\right) = \begin{bmatrix} e^a & 0 \\ 0 & e^a \end{bmatrix}\begin{bmatrix} \cos b & \sin b \\ -\sin b & \cos b \end{bmatrix} \in \mathbb{C} \qquad (6.7)$$

The minus sign is calculated by insisting upon multiplicative closure of matrix form[46].

Division algebras, spinor algebras, and with them the associated spinor spaces, emerge similarly from every finite group. Our area of interest in this book is the spinors which emerge from the finite groups, C_2 , $C_2 \times C_2$, and $C_2 \times C_2 \times C_2$; we will also look briefly at the spinors in the $C_2 \times C_2 \times C_2 \times C_2$ group. These spinors correspond to the Clifford algebras of the appropriate dimensions. It is within these spinor spaces that we find chirality. We remind the reader not to presume that any commutative finite group or any non-commutative finite group holds chirality. Chirality is a rare jewel.

Angles and rotation in spinor spaces:
Spinor spaces contain a rotation matrix and within that rotation matrix a set of trigonometric functions. These trigonometric functions have an argument comprised of $N-1$ real variables where N is the order of the finite group underlying the spinor space – see (6.1). The argument of the trigonometric functions is the angle of the spinor space.

Each spinor space has only one angle, but it is a multi-dimensional angle. Each of the 3-dimensional spinor spaces has a 3-dimensional angle comprised of two real variables; each of the 4-dimensional spinor spaces, like the quaternions, have a 4-dimensional angle comprised of three real variables, and so on. This is significantly different from the $\mathbb{R}^n$ spaces with which we are experientially familiar and which hold only 2-dimensional angles which are each a single real variable.

In our 4-dimensional space-time, rotation is 2-dimensional. Two of the four dimensions of our space-time are unaffected by 2-dimensional rotation. If the rotation is Euclidean, rather than a space-time rotation, one of these unaffected dimensions is the time dimension and the other unaffected dimension is a spatial dimension. We call the unaffected spatial dimension the axis of our spatial rotation, and so we have come to believe that rotation is about a single spatial axis. In our 4-dimensional space-time, 2-dimensional rotation is about two axes, but wholly spatial rotation is rotation in which only one of these two axes is a spatial axis; this is the genesis of our belief that rotation is about an axis.

Rotation in a spinor space is not rotation about an axis as taking the eigenvectors of any spinor rotation matrix will prove – every eigenvector is dependent upon the angle. Rotation in a spinor space is about a point. The complex plane, $\mathbb{C}$, gives an example of spinor rotation. There are only two dimensions in the complex plane; there is no third dimension to be an axis of rotation.

---

[46] See: Dennis Morris : The Physics of Empty Space.

Distance in spinor spaces:

A spinor space has a distance function which is the norm of the associated division algebra. The distance function of a spinor space is the determinant of the algebraic matrix form of that space. These distance functions are not 'just invented' by mathematicians; these distance functions are derived directly from the finite groups.

An important point:

The complex numbers, $\mathbb{C}$, are a division algebra for all values of the real variable. The hyperbolic complex numbers are a division algebra for only positive values of the real variable.

The determinant of the algebraic matrix form of the hyperbolic complex numbers, (6.5), is:

$$\det\left(\begin{bmatrix} t & z \\ z & t \end{bmatrix}\right) = t^2 - z^2 \tag{6.8}$$

This determinant can be zero implying the algebraic matrix form can be singular. Singular matrices do not have an inverse, and so they cannot be elements of a division algebra. Such singular matrices are automatically excluded by taking the exponential of the matrix. Thus the hyperbolic complex numbers are the polar form only, (6.6). We see in (6.6) that the real variable, e^a, is always positive; the imaginary variable can be positive or negative because the determinant of a rotation matrix is always unity[47].

The vast majority of spinor algebras are algebras in only the polar form. The two types of quaternions and the Euclidean complex numbers are the only spinor algebras in which the real variable can take negative values[48]. The quaternions, $\mathbb{H}$, like the complex numbers, $\mathbb{C}$, are a division algebra for all values of the real variable. The determinant of the complex number algebraic matrix form, $\mathbb{C}$, and the determinant of the quaternion algebraic matrix form, $\mathbb{H}$, are:

$$\det(\mathbb{C}) = x^2 + y^2$$
$$\det(\mathbb{H}) = \left(w^2 + x^2 + y^2 + z^2\right)^2 \tag{6.9}$$

These determinants can never be zero unless all variables are zero.

Conserved charges in spinor spaces:

Consequent to Noether's theorem, when each spinor space is taken to be a gauge space over our 4-dimensional space-time, each type of rotation in that spinor space has associated with it a conserved 'charge' which we take to be the real variable of the spinor algebra. The imaginary variables are taken to be a conserved 'current'. In this book, the real quaternion variable is taken to be electric charge, and so, because the quaternion real variable can be either positive or negative, we have both positive and negative electric charge in the universe. The imaginary quaternion variables are taken to be intrinsic angular momentum. The 4-dimensional A_3 algebras are like the hyperbolic complex numbers in that each is a division algebra for only positive values of the real variable. In this book, the real A_3 variable, the A_3

---

[47] The determinant of the exponential of a matrix with zero trace is always unity.
[48] Amazingly, some 4-dimensional sub-algebras of 8-dimensional Clifford algebras can have negative real variables.

charge associated with the space-time part of the rotation in the A_3 space, is taken to be mass, and so we have only positive mass.

We will later present the idea that anti-matter exists in our universe because the quaternion real variable can be either positive or negative. We will later present the idea that there is no negative matter within our universe because the A_3 real variable associated with the space-time part of the A_3 rotation can be only positive.

Actually, in more detail, we will associate the symmetric imaginary A_3 variables with momentum and we will associate the anti-symmetric imaginary A_3 variables with classical magnetism. The real variable in the A_3 algebras will thus have a dual role; in the anti-symmetric case the real variable will be associated with classical electric charge; in the symmetric case the real variable will be associated with mass. The complications in the A_3 algebras arise because there are both symmetric and anti-symmetric variables within them; there are only anti-symmetric variables in the two quaternion algebras. The symmetric variables are associated with a space-time type of rotation and the anti-symmetric variables are associated with a euclidean type of rotation; thus there are two types of conserved charge in the A_3 spaces. Note that there is no classical positive electric field and that classical magnetism rotates in only one direction.

Virtual processes – off mass shell:

Within our 4-dimensional space-time, mass/energy is conserved and momentum is conserved because these are a conserved charge and a conserved current of the A_3 algebras, the A_3 spinor spaces. In quaternion space, electric charge and intrinsic angular momentum are conserved because these are the conserved charge and conserved current of the quaternion algebras, the quaternion spinor spaces.

Does this mean that processes can happen in quaternion space which are not constrained to conserve momentum or energy but do conserve electric charge and intrinsic spin? Have such processes ever been observed? Such processes which violate momentum conservation and energy conservation have never been observed in our 4-dimensional space-time, but they are presumed to occur.

Physicists calculate the probable outcome of experiments in quantum field theory using Feynman diagrams which are graphical representations of what are called 'virtual processes'. The virtual processes violate energy conservation, and they violate momentum conservation. The virtual processes are said to be 'off mass shell' because they violate energy conservation and they violate momentum conservation. By this we mean that, in a virtual process, we have:

$$\underline{\textit{Off mass shell}}$$

$$\textit{Massive particles}: \qquad E^2 \neq p^2 c^2 + m^2 c^4 \qquad\qquad (6.10)$$
$$\textit{Massless particles}: \qquad E \neq pc$$

The reality of virtual processes is utterly central to quantum field theory. Without virtual processes, we have no understanding of quantum field theory. It is the virtual processes that are the source of probability within quantum physics.

We argue that, if energy and momentum are conserved within our 4-dimensional space-time, then any process which violates these conservation laws must be outside of our 4-dimensional space-time. We take the view that 'off mass shell' events happen in quaternion space and that these events happen outside of our 4-dimensional space-time. The reader will form their own opinion.

Derivatives in spinor spaces:

A spinor space has a derivative – see later note. Within spinor space, a potential, Φ, is an element of the spinor algebra, a spinor. Each of the individual spinor variables in a potential, one real variable and several imaginary variables, are functions of several variables[49]. We non-commutatively differentiate this potential with a differential operator[50], d, acting as if by matrix multiplication to get an E-field and a B-field[51]:

$$E = \frac{1}{2}(d\Phi + \Phi d)$$
$$B = \frac{1}{2}(d\Phi - \Phi d)$$

(6.11)

When the division algebra of the spinor space is non-commutative, like quaternions, both the E-field and the B-field are non-zero. When the division algebra is commutative, the B-field is zero and the E-field reduces to the conventional derivative.

The anti-symmetric part of the B-field corresponds to the classical magnetic field in the A_3 algebras, and the B-field corresponds to the neutrino field in the quaternion algebras. The anti-symmetric part of the E-field corresponds to the classical electric field in the A_3 algebras, and the E-field corresponds to the electron field in the quaternion algebras. The symmetric parts of the E-field and the B-field of the A_3 algebras correspond to the energy-momentum tensor of general relativity.

In a later chapters, we will see that the B-field is chiral. We will associate the B-field with the chirality of the spinor space.

Emergent spaces:

The spinor spaces are the building blocks, angles, trigonometric functions, rotation matrices, E-fields, B-fields, and distance functions, of our 4-dimensional space-time. When the mathematical conditions are right, as they are only once[52], the spinor space building blocks are fabricated together to form a non-algebraic space which we call an emergent space. There is only one emergent space; it is our 4-dimensional space-time. There are semi-emergent spaces which are almost a non-algebraic space; the quaternion emergent space is one. These semi-emergent spaces seem to play a role as gauge spaces within our universe.

---

[49] Each function within a spinor potential is a function of as many variables as there are variables in the spinor algebra.
[50] We just use the differential operator as a calculational aid. We can differentiate without the operator if we choose.
[51] For details of non-commutative differentiation, see : Dennis Morris : The Physics of Empty Space.
[52] See : Dennis Morris : The Uniqueness of our Space-time.

Clifford algebras:

Although there are an infinite number of spinor spaces corresponding to the infinite number of finite groups, we are interested in only a few of these spinor spaces. As we said before, our interest is in the spinor spaces that derive from the finite groups C_2, $C_2 \times C_2$, and $C_2 \times C_2 \times C_2$. These spinor spaces correspond to the 2-dimensional[53], 4-dimensional, and 8-dimensional Clifford algebras. We give a brief introduction to the Clifford algebras in a later chapter.

---

[53] Traditionally, the 2-dimensional algebras are not seen as being Clifford algebras because they are commutative algebras.

Chapter 7

A Note upon Matrix Differentiation

Although matrix differentiation is covered elsewhere[54], we present a few notes to remind the reader.

The differential of a 2-dimensional complex number field, $\mathbb{C}$, is:

$$d\begin{bmatrix} f(x,y) & g(x,y) \\ -g(x,y) & f(x,y) \end{bmatrix} = \frac{\partial\begin{bmatrix} f & g \\ -g & f \end{bmatrix}}{\partial\begin{bmatrix} x & 0 \\ 0 & x \end{bmatrix}} + \frac{\partial\begin{bmatrix} f & g \\ -g & f \end{bmatrix}}{\partial\begin{bmatrix} 0 & y \\ -y & 0 \end{bmatrix}}$$

$$= \frac{\partial\begin{bmatrix} f & g \\ -g & f \end{bmatrix}}{\partial\begin{bmatrix} x & 0 \\ 0 & x \end{bmatrix}} + \begin{bmatrix} 0 & 1 \\ -1 & 0 \end{bmatrix}\frac{\partial\begin{bmatrix} f & g \\ -g & f \end{bmatrix}}{\partial\begin{bmatrix} y & 0 \\ 0 & y \end{bmatrix}}$$

$$= \begin{bmatrix} \dfrac{\partial f}{\partial x} + \dfrac{\partial g}{\partial y} & \dfrac{\partial g}{\partial x} - \dfrac{\partial f}{\partial y} \\ -\left(\dfrac{\partial g}{\partial x} - \dfrac{\partial f}{\partial y}\right) & \dfrac{\partial f}{\partial x} + \dfrac{\partial g}{\partial y} \end{bmatrix}$$

$$= \begin{bmatrix} Div & Curl \\ -Curl & Div \end{bmatrix} \tag{7.1}$$

The differential operator:

The differential operator is a calculation convenience which allows us to differentiate matrices by 'matrix multiplication'. This works because both matrix multiplication and differentiation are linear operations. The differential operator is formed as a sum of the inverse of each variable within the algebraic matrix form. We have to use the inverses because, see above (7.1), the differential variables occur as denominators. Note that the form of the differential operator, which is the sum of the inverses of each individual variable, is not the inverse of the algebraic matrix form.

Pretending that differentiation is matrix multiplication, we have:

$$\begin{bmatrix} \partial x & -\partial y \\ \partial y & \partial x \end{bmatrix}\begin{bmatrix} f & g \\ -g & f \end{bmatrix} = \begin{bmatrix} \dfrac{\partial f}{\partial x} + \dfrac{\partial g}{\partial y} & \dfrac{\partial g}{\partial x} - \dfrac{\partial f}{\partial y} \\ -\left(\dfrac{\partial g}{\partial x} - \dfrac{\partial f}{\partial y}\right) & \dfrac{\partial f}{\partial x} + \dfrac{\partial g}{\partial y} \end{bmatrix} \tag{7.2}$$

---

[54] See : Dennis Morris : Complex Numbers The Higher Dimensional Forms – 2nd edition.

Because the complex numbers, $\mathbb{C}$, are commutative, it does not matter from which side the differential operator acts.

In the case of the 2-dimensional hyperbolic complex numbers, $\mathbb{S}$, which is 2-dimensional space-time, we have:

$$
\begin{bmatrix} \partial t & \partial z \\ \partial z & \partial t \end{bmatrix} \begin{bmatrix} f & g \\ g & f \end{bmatrix} = \begin{bmatrix} \dfrac{\partial f}{\partial t} + \dfrac{\partial g}{\partial z} & \dfrac{\partial g}{\partial t} + \dfrac{\partial f}{\partial z} \\ \dfrac{\partial g}{\partial t} + \dfrac{\partial f}{\partial z} & \dfrac{\partial f}{\partial t} + \dfrac{\partial g}{\partial z} \end{bmatrix} = \begin{bmatrix} Div_{ST} & Curl_{ST} \\ Curl_{ST} & Div_{ST} \end{bmatrix}
\tag{7.3}
$$

The curls:

The curl of a field is of a rotational nature. The position of the curl within the matrix is such that the exponential of that matrix with zero divergence is a rotation matrix:

$$
\exp\left(\begin{bmatrix} 0 & C \\ C & 0 \end{bmatrix} \right) = \begin{bmatrix} \cosh C & \sinh C \\ \sinh C & \cosh C \end{bmatrix}
\tag{7.4}
$$

Within 2-dimensional Euclidean space, the curl is:

$$
Curl = \frac{\partial g}{\partial x} - \frac{\partial f}{\partial y}
\tag{7.5}
$$

Within 2-dimensional space-time, the curl is:

$$
Curl_{ST} = \frac{\partial g}{\partial t} + \frac{\partial f}{\partial z}
\tag{7.6}
$$

Note the sign difference between (7.5) & (7.6).

Within 2-dimensional space-time, rotation is change of velocity, acceleration, whereas in 2-dimensional Euclidean space, rotation is, well, rotation.

The Euclidean curl is associated with a spatially rotating field. The 2-dimensional space-time curl is associated with linear acceleration – an accelerating force field. We see this in the classical electric field which is of the form of a space-time curl (both signs the same):

$$
E_x = -\frac{\partial \phi}{\partial x} - \frac{\partial A_x}{\partial t}
$$
$$
E_y = -\frac{\partial \phi}{\partial y} - \frac{\partial A_y}{\partial t}
\tag{7.7}
$$
$$
E_z = -\frac{\partial \phi}{\partial z} - \frac{\partial A_z}{\partial t}
$$

Note that we must have the time variable within a space-time curl – obvious really.

The signs are arbitrarily reversed by convention. Of course, an electric field accelerates a charged particle in a particular direction. A classical magnetic field is circular and is of the form of a spatial curl:

$$B_x = \frac{\partial A_z}{\partial y} - \frac{\partial A_y}{\partial z}$$

$$B_y = \frac{\partial A_x}{\partial z} - \frac{\partial A_z}{\partial x} \tag{7.8}$$

$$B_z = \frac{\partial A_y}{\partial x} - \frac{\partial A_x}{\partial y}$$

Note that we must not have the time variable in a wholly spatial curl – also obvious really.

If, as we will, we get space-time curls occurring in our derivatives, we will know them to correspond to linear acceleration fields like the electric field. If, as we will, we get Euclidean curls occurring in our derivatives, we will know them to correspond to rotational fields like the magnetic field.

An important point:

Our 4-dimensional space-time is a fabrication of six 2-dimensional spinor spaces. As such, only 2-dimensional curls can exist in our 4-dimensional space-time. If our deliberations lead to some form of expression for a curl other than the forms of the 2-dimensional curls, (7.5) & (7.6), then we opine that this 'other expression curl' will not be manifest in our 4-dimensional space-time.

Non-commutative differentiation:

Non-commutative differentiation is covered in detail elsewhere[55].

Since matrices are not always multiplicatively commutative, differentiation by a differential operator is not always commutative. An example is differentiation within a quaternion algebra. The left-chiral quaternion differential operator[56] is:

$$d_{\mathbb{H}_{L\chi}} = \begin{bmatrix} \partial t & -\partial x & -\partial y & -\partial z \\ \partial x & \partial t & \partial z & -\partial y \\ \partial y & -\partial z & \partial t & \partial x \\ \partial z & \partial y & -\partial x & \partial t \end{bmatrix} \tag{7.9}$$

This operator acts as if by matrix multiplication upon a left-chiral quaternion potential:

$$\Phi_{\mathbb{H}_{L\chi}} = \begin{bmatrix} \phi & -A_x & -A_y & -A_z \\ A_x & \phi & A_z & -A_y \\ A_y & -A_z & \phi & A_x \\ A_z & A_y & -A_x & \phi \end{bmatrix} \qquad : \qquad \begin{array}{l} \phi(t,x,y,z) \\ A_x(t,x,y,z) \\ A_y(t,x,y,z) \\ A_z(t,x,y,z) \end{array} \tag{7.10}$$

---

[55] See : Dennis Morris : The Physics of Empty Space.
[56] Let us be accurate. There is no such thing as a differential operator. We have differentiation which really exists. The differential 'operator' is no more than a calculative aid which enables us to differentiate easily. All that we do with the differential operator could be equally well done 'properly'. Do not worship operators; they do not exist.

This operator acts non-commutatively to both the left and the right of the potential to produce both an E-field and a B-field[57]:

$$E = \frac{1}{2}(d\Phi + \Phi d)$$
$$B = \frac{1}{2}(d\Phi - \Phi d)$$

(7.11)

We see that we get two fields as differentials of a non-commutative potential. We will see many examples of non-commutative differentiation shortly.

We see that the B-field reduces to zero if the field is commutative. Note that this non-commutative differentiation makes sense only if the non-commutative nature of the field is such that $BC = -CB$. In other words, we can differentiate non-commutatively in only the chiral algebras. If forces, in the form of curls, are necessarily the differentials of potentials, then we can have forces from only the chiral algebras because these are the only algebras which have non-commutative differentials. We see that our list of possible forces grows thin; it is not zero, but it is thin – this fits with observation.

That's such a good point, I think I will repeat it;

  a) we can differentiate non-commutatively in only the chiral algebras.
  b) we can have forces which are differentials of potentials from only the chiral algebras.

There is a further point. The differentials must produce curls of a 2-dimensional nature for these curls to be manifest in our 4-dimensional space-time because our 4-dimensional space-time holds only 2-dimensional rotations. For example, the differentials of the chiral algebras of the $C_2 \times C_6$ group cannot be manifest in our 4-dimensional space-time because they are of the wrong form.

---

[57] See: Dennis Morris : The Physics of Empty Space.

Chapter 8

An Overview of Traditional Clifford Algebra

In this chapter, we tie up one more loose end. It is not essential that the reader digests this material to understand the remainder of the book. The material in this chapter is presented in more detail and more extensively in the book 'The Naked Spinor' by Dennis Morris.

Clifford algebras:

The Clifford algebras were formulated by William Kingdon Clifford (1845-1879) in 1876 and independently by Rudolf Lipshitz[58] (1832-1903) in 1880. Clifford's work was published posthumously in 1882. When Clifford did his work, the concept of vectors had recently been formulated by Hermann Grassmann (1809-1877), and Clifford was seeking a way to multiply together two vectors which was both associative and distributive.

Clifford took the view that all types of space have a distance function which is a quadratic form like:

$$dist^2 = \pm x^2 \pm y^2$$
$$dist^2 = \pm t^2 \pm x^2 \pm y^2 \pm z^2$$

(8.1)

Note that such spaces can hold as many 2-dimensional rotations as there are pairs of variables.

Clifford's view is a common enough view even among mathematicians today; it was certainly the view of the founder of Riemann geometry Bernhard Riemann (1826-1866), and it is certainly the observed nature of our 4-dimensional space-time. Associated with each of the variables in a distance function is a basis vector pointing in the spatial direction associated with the particular variable:

$$\vec{e}_x, \quad \vec{e}_y, \quad \dots$$

(8.2)

Vectors are formed as sums of real multiples of these basis vectors. The sum of two such real multiples of basis vectors is a 'hypotenuse' vector:

$$x\vec{e}_x + y\vec{e}_y$$

(8.3)

We know by Pythagoras that, in Euclidean space, the squared length of this hypotenuse is $x^2 + y^2$. If we use 'proper' multiplication to multiply vectors together, then this multiplication operation must be within a division algebra. Within a division algebra, we must have 'product of norms equals norm of products'; we should therefore have:

$$\left(x\vec{e}_x + y\vec{e}_y\right)\left(x\vec{e}_x + y\vec{e}_y\right) = x^2 + y^2$$

(8.4)

---

[58] See : R. Lipshitz : Untersuchungen uber die Summen von Quadraten - 1886

Okay, the logic behind this, (8.4), might be questionable, but this is the way Clifford went; let us continue. We will assume that multiplication of two vectors has the same associative and distributive rules as algebraic multiplication – it is 'proper multiplication', but we will be careful to not assume multiplicative commutativity. We change the notation slightly by changing the subscripts of the basis vectors:

$$\left(x\vec{e_1}+y\vec{e_2}\right)\left(x\vec{e_1}+y\vec{e_2}\right)=x^2\vec{e_1}\vec{e_1}+xy\vec{e_1}\vec{e_2}+yx\vec{e_2}\vec{e_1}+y^2\vec{e_2}\vec{e_2} \tag{8.5}$$

To arrive at the required result, (8.4), we need:

$$\vec{e_1}\vec{e_1}=1, \qquad \vec{e_2}\vec{e_2}=1, \qquad \vec{e_1}\vec{e_2}=-\vec{e_2}\vec{e_1}$$
$$\left(\vec{e_1}\vec{e_2}\right)\left(\vec{e_1}\vec{e_2}\right)=\vec{e_1}\vec{e_2}\vec{e_1}\vec{e_2}=-\vec{e_1}\vec{e_1}\vec{e_2}\vec{e_2}=-1 \tag{8.6}$$

Looking at the $\vec{e_1}\vec{e_2}$ in (8.6), we see the chiral kind of non-commutativity.

Bi-vectors and tri-vectors:

The pairings of basis vectors like $\vec{e_1}\vec{e_2}$ are called bi-vectors. Three such basis vectors, $\vec{e_1}\vec{e_2}\vec{e_3}$, would be called a tri-vector. They are often written with only one full-size script and several subscripts like $\vec{e_1}\vec{e_2}=\vec{e}_{12}$ or $\vec{e_1}\vec{e_3}\vec{e_2}=\vec{e}_{132}$.

At this point, although Clifford's logic behind the above might be questionable, the mathematics is starting to make sense. Unfortunately, at this point, Clifford algebraists go off at a tangent and start talking about geometry; they claim a vector is a length, a bi-vector is an area, a tri-vector is a volume etc.. This geometric distraction is, in your author's opinion[59], best forgotten.

Returning to (8.6), we have a real number (the 1), two basis vectors, and a basis bi-vector. We can present this Clifford algebra as the four elements:

$$a.1+b.\vec{e_1}+c.\vec{e_2}+d.\vec{e_1}\vec{e_2}$$
$$1, \qquad \vec{e_1}=\sqrt{+1}, \qquad \vec{e_2}=\sqrt{+1}, \qquad \vec{e_1}\vec{e_2}=\sqrt{-1} \tag{8.7}$$

This, (8.7), is a 4-dimensional Clifford algebra; it is called the Clifford algebra $Cl_{2,0}$. The 2 refers to two plus signs in the distance function which generated it, (8.4); the zero refers to zero minus signs in the distance function which generated it, (8.4). Presumably, the Cl refers to a marine mammal of some nature.

We see that we have something like a 4-dimensional complex number. There is a problem; this, (8.7), construction contains zero divisors[60], and so it cannot be a division algebra with 'proper' multiplication; this is true of all constructions which contain a real variable and a square root of plus unity for in such cases we always have the zero product of two non-zero elements of the algebra:

$$\left(1+\sqrt{+1}\right)\left(1-\sqrt{+1}\right)=0 \tag{8.8}$$

---

[59] I do not doubt there are many highly competent mathematicians who might hold a different opinion.
[60] Zero divisors are two non-zero elements of the algebra whose product is zero.

A non-traditional approach:

Instead of writing the different variables within a space as basis vectors with little arrows over them, let us write them as (mutually orthogonal) permutation matrices[61].

Now, symmetric permutation matrices are square roots of plus unity and 'anti-symmetric permutation matrices'[62] are square roots of minus unity. Looking at (8.7), you can see where we are going. We have:

$$a.1 + b.\vec{e_1} + c.\vec{e_2} + d.\vec{e_1}\vec{e_2} \equiv \begin{bmatrix} a & b & c & d \\ b & a & d & c \\ c & -d & a & -b \\ -d & c & -b & a \end{bmatrix} \tag{8.9}$$

Note that we have now dispensed with the concepts of basis vectors and basis bi-vectors; instead, we have orthogonal variables. This matrix has all the algebraic properties of (8.7), but it gets better. Taking the exponential of the above matrix, (8.9), rids us of the zero divisors and the singular matrices. By taking the exponential, we turn the Clifford algebra $Cl_{2,0}$ into a division[63] algebra[64]. In this case, (8.9), we have one of the six A_3 division algebras:

$$A_{3L\chi}^{SSA} = \exp\left(\begin{bmatrix} a & b & c & d \\ b & a & d & c \\ c & -d & a & -b \\ -d & c & -b & a \end{bmatrix}\right) \tag{8.10}$$

The other 4-dimensional Clifford algebras:

There are two other 4-dimensional Clifford algebras based upon the distance functions:

$$\begin{aligned} dist^2 &= x^2 - y^2 \\ dist^2 &= -x^2 - y^2 \end{aligned} \tag{8.11}$$

These Clifford algebras are known as $Cl_{1,1}$ & $Cl_{0,2}$, reflecting the numbers of plus and minus signs in the distance functions. As division algebras, the two Clifford algebras $Cl_{1,1}$ & $Cl_{2,0}$ are isomorphic – they

---

[61] A permutation matrix is a square matrix with a single 1 in each row and a single 1 in each column and zeros as every other element. There is a one-to-one correspondence between permutation matrices and permutations. Closed sets of mutually orthogonal permutation matrices are finite groups. The elements of finite groups are mutually orthogonal in the sense that they are not 'mixed together'.

[62] Technically, permutation matrices do not include minus signs and so we cannot have anti-symmetric permutation matrices, and so we misuse the term permutation matrix.

[63] A division algebra is a type of complex numbers. The space associated with a division algebra, like the complex plane, is often called a spinor space.

[64] This is a division algebra without additive inverses on the real axis or an additive identity on the real axis – like space-time.

are both copies of an A_3 algebra. The $Cl_{0,2}$ Clifford algebra has three square roots of minus unity – it is a quaternion.

The finite group connection:

The six A_3 algebras and the two quaternion algebras derive from the finite group $C_2 \times C_2$. The two 2-dimensional 'Clifford algebras' $\mathbb{C} \& \mathbb{S}$ derive from the finite group C_2. The 8-dimensional Clifford algebras derive from the finite group $C_2 \times C_2 \times C_2$. The 16-dimensional Clifford algebras derive from the finite group $C_2 \times C_2 \times C_2 \times C_2$.

Higher dimensional Clifford algebras:

Higher dimensional Clifford algebras were traditionally found by using higher dimensional distance functions; for example, we require:

$$\left(x\vec{e_1} + y\vec{e_2} + z\vec{e_3} \right)\left(x\vec{e_1} + y\vec{e_2} + z\vec{e_3} \right) = x^2 + y^2 + z^2 \tag{8.12}$$

We have:

$$\left(x\vec{e_1} + y\vec{e_2} + z\vec{e_3} \right)\left(x\vec{e_1} + y\vec{e_2} + z\vec{e_3} \right) =$$
$$x^2\vec{e_1}\vec{e_1} + xy\vec{e_1}\vec{e_2} + xz\vec{e_1}\vec{e_3} + yy\vec{e_2}\vec{e_1} + y^2\vec{e_2}\vec{e_2}$$
$$+ yz\vec{e_2}\vec{e_3} + xz\vec{e_3}\vec{e_1} + yz\vec{e_3}\vec{e_2} + z^2\vec{e_3}\vec{e_3} \tag{8.13}$$

This, (8.12) & (8.13), leads to:

$$\vec{e_1}\vec{e_1} = \vec{e_2}\vec{e_2} = \vec{e_3}\vec{e_3} = +1$$
$$\vec{e_1}\vec{e_2} = -\vec{e_2}\vec{e_1}, \quad \vec{e_1}\vec{e_3} = -\vec{e_3}\vec{e_1}, \quad \vec{e_2}\vec{e_3} = -\vec{e_3}\vec{e_2} \tag{8.14}$$

$$\left(\vec{e_1}\vec{e_2} \right)^2 = \left(\vec{e_1}\vec{e_3} \right)^2 = \left(\vec{e_2}\vec{e_3} \right)^2 = -1$$
$$\left(\vec{e_1}\vec{e_2}\vec{e_3} \right)^2 = -\vec{e_1}\vec{e_1}\vec{e_2}\vec{e_2}\vec{e_3}\vec{e_3} = -1 \tag{8.15}$$

This algebra is known as $Cl_{3,0}$. We can characterise this algebra as[65]:

$$Cl_{3,0} = a + b\vec{e_1} + c\vec{e_2} + d\vec{e_3} + e\vec{e_{12}} + f\vec{e_{13}} + g\vec{e_{23}} + h\vec{e_{123}} \tag{8.16}$$

This, (8.16), is an 8-dimensional Clifford algebra. This can be presented as a 8×8 matrix analogously to (8.9). We see there are three square roots of plus unity and four square roots of minus unity; that is three symmetric 8×8 permutation matrices and four anti-symmetric 8×8 permutation matrices.

---

[65] I apologise for the notational clash of the e's.

Non-commutativity:

Using the above relations, (8.15), we see the non-commutative nature of bi-vectors:

$$\overrightarrow{e_{12}}\overrightarrow{e_{23}} = \overrightarrow{e_1}\left(\overrightarrow{e_2 e_2}\right)\overrightarrow{e_3} = \overrightarrow{e_1 e_3} = \overrightarrow{e_{13}}$$
$$\overrightarrow{e_{23}}\overrightarrow{e_{12}} = \overrightarrow{e_2 e_3 e_1 e_2} = -\overrightarrow{e_2 e_3 e_2 e_1} = +\overrightarrow{e_2 e_2 e_3 e_1} = \overrightarrow{e_3 e_1} = -\overrightarrow{e_1 e_3} = -\overrightarrow{e_{13}}$$

(8.17)

Remarkably, as the reader will find if she plays with the above relations, (8.15), the tri-vector $\overrightarrow{e_1 e_2 e_3}$ commutes with all other elements of the algebra; for example:

$$\overrightarrow{e_{12}}\overrightarrow{e_{123}} = \overrightarrow{e_1 e_2 e_1 e_2 e_3} = -\overrightarrow{e_1 e_1 e_2 e_2 e_3} = -\left(\overrightarrow{e_1 e_1}\right)\left(\overrightarrow{e_2 e_2}\right)\overrightarrow{e_3} = -\overrightarrow{e_3}$$
$$\overrightarrow{e_{123}}\overrightarrow{e_{12}} = \overrightarrow{e_1 e_2 e_3 e_1 e_2} = +\overrightarrow{e_1 e_2 e_2 e_3 e_1} = -\overrightarrow{e_2 e_2 e_3 e_1 e_1} = -\overrightarrow{e_3}$$

(8.18)

In general, the tri-vector of every 8-dimensional Clifford algebra commutes with all other elements of the algebra. The structure of every 8-dimensional Clifford algebra is such that it has two commutative elements, the real number and the tri-vector, and six non-commutative elements, the vectors and the bi-vectors, – that's worth remembering.

Permutation matrices and groups (a little repetition):

Above, (8.9), we formed the matrix representation of a Clifford algebra using permutation matrices. We can do the same with the 8-dimensional Clifford algebras and, indeed, with every Clifford algebra.

Finite groups are no more than closed sets of permutations. A closed set of permutations can be written as a closed set of permutation matrices. Matrix multiplication of permutation matrices is no-more than sequential combination of permutations. In short, each of the Clifford algebras is underlain by a finite group.

Denoting the Clifford algebras:

Traditionally, Clifford algebras are denoted by $Cl_{i,j}$. We find it convenient to also denote them by a list of the algebraic nature of each of the basis elements of the algebra. All Clifford algebras have basis elements which are square roots of plus unity, $\sqrt{+1}$, or square roots of minus unity, $\sqrt{-1}$, and so we can specify the particular Clifford algebra by the number of each type of basis element; an example is:

$$Cl_{0,3} = 1 + 1\sqrt{+1} + 6\sqrt{-1}$$

(8.19)

It is important to realise that this means of denoting the Clifford algebras does not imply a Clifford algebra actually has a square root of plus unity; we need to take the exponential to form the algebra. Perhaps we ought to denote the Clifford algebras as:

$$Cl_{0,3} = \exp\left(1 + 1\sqrt{+1} + 6\sqrt{-1}\right)$$

(8.20)

The denotation without the exponential has become standard.

The 8-dimensional Clifford algebras:

Although there are four separate 8-dimensional Clifford algebras based upon the four possible combinations of minus signs and plus signs in the three variables distance function, these four Clifford algebras are only three separate non-commutative 8-dimensional division algebras.

$$Cl_{3,0} \cong Cl_{1,2} = 1 + 3\sqrt{+1} + 4\sqrt{-1}_{Non-Com}$$

$$Cl_{2,1} = 1 + 5\sqrt{+1} + 2\sqrt{-1} \qquad (8.21)$$

$$Cl_{0,3} = 1 + 1\sqrt{+1} + 6\sqrt{-1}$$

Summary:

We have seen that the Clifford algebras have the $BC = -CB$ kind of non-commutativity. At first, it might seem that the Clifford algebras are a third mathematical source of this $BC = -CB$ kind of non-commutativity, but we have seen that the Clifford algebras are really the non-commutative algebras which derive from the $C_2 \times C_2 \times ...$ type of finite groups.

Parity and CP Invariance

It is a basic and easily verifiable fact that a physical apparatus of any form works exactly the same when it is pointing north as when it is pointing west. Chemical reactions in test tubes work exactly the same regardless of whether the test tube is pointing east or pointing south. Car engines work just as efficiently when the car is travelling north as when the car is travelling south. We say that the physics of the universe is invariant under rotation in 2-dimensional Euclidean space. We might equally well say that the universe is isotropic.

It is a basic fact, but not so easily verifiable, that a physical apparatus of any form works exactly the same when it is moving at 100,000 miles per second as it does when it is stationary. Chemical reactions in test tubes work exactly the same regardless of whether the test tube is moving at 50,000 miles per second or moving at 10 miles per hour. Car engines work just as efficiently when the car is in a rocket moving at 10,000 miles per second as when the car is stationary. We say that the physics of the universe is invariant under change of velocity. This is the special theory of relativity, of course. Change of velocity is just rotation in 2-dimensional space-time. We can say that the physics of the universe is invariant under rotation in 2-dimensional space-time.

Given that the physics of the universe is invariant under rotation, we might expect that the physics of the universe would be invariant under spatial reflection. In our 4-dimensional space-time, spatial reflection is called spatial parity. In mathematical terms, spatial reflection, spatial parity, is simply reversing the sign of the spatial components of the mathematical expression describing the physical system. Physicists have the 'spatial parity operator', P, which exactly reverses the sign of the spatial components of a system:

$$P\big(\psi\left(t,x,y,z\right)\big)=\psi\left(t,-x,-y,-z\right)$$
$$P^2\big(\psi\left(t,x,y,z\right)\big)=\psi\left(t,x,y,z\right)$$

(9.1)

If the spatial parity operator is applied twice, we get back to where we started.

We take the view that, within our 4-dimensional space-time, spatial parity is inviolable. Our view is that spatial parity is a fundamentally intrinsic property of our 4-dimensional space-time. If spatial parity is violated, then that violation must happen outside of our 4-dimensional space-time.

Spatial parity and chirality:
Spatial parity is often taken to be the same as chirality, but, in this book, we will take the view that spatial parity is a different thing from chirality. We will take the view that chirality is a general term used in all types of space whereas spatial parity is a specific term used in only our 4-dimensional space-time. The chirality of our 4-dimensional space-time is spatial parity. Other types of space have their own kind of chirality.

Intrinsic parity:

Rather confusingly, there is a quantum number associated with sub-atomic particles called intrinsic parity. The reader might have noticed that above we have referred to 'spatial parity'. Intrinsic parity is a different thing from spatial parity. Unfortunately, many texts refer simply to parity, and so these two concepts become confused.

Subatomic particles like the pion decay into other particles. For example, the neutral pion most often, 98.8% of the time, decays into two photons:

$$\pi^0 \rightarrow \gamma\gamma \qquad\qquad (9.2)$$

We never see a pion decay into one photon or into three photons or into any odd number of photons. Why does the pion not decay into three photons? We assume there is a conserved multiplicative quantum number, which we call intrinsic parity, such that the photon has intrinsic parity of -1. The total intrinsic parity of a pair of photons is the product of the intrinsic parity of each photon $(-1)\times(-1) = +1$. We say the pion has an intrinsic parity of $+1$. The pion cannot decay into three photons because the product of the intrinsic parities of three photons is -1. We need to balance the intrinsic parity on either side of the reaction, (9.2). Intrinsic parity is invented by humankind to explain the absence of many otherwise seemingly reasonable particle decay routes.

Photons and gluons both have intrinsic parity of -1. Electrons, neutrinos and quarks all have intrinsic parity of $+1$. The anti-particles of electrons, neutrinos and quarks which are positrons, anti-neutrinos, and anti-quarks have intrinsic parity of $+1$. In general, the conventional understanding of anti-matter is such that anti-particles have opposite intrinsic parity to the corresponding particles, but things are not that clear. The weak force bosons, W^+, W^-, Z^0 do not have a definite intrinsic parity – they are not eigenstates of the intrinsic parity operator, and neutral pions are often thought of as being their own anti-particles.

In this book, we do not dwell upon intrinsic parity. We simply wanted to clarify the meaning of the word 'parity' for the reader.

Spatial parity and the violation of spatial parity:

Spatial parity is the type of chirality with which the reader will be most familiar. The reader might have heard that the weak nuclear force violates parity – this is spatial parity. Spatial parity, often called conservation of parity, is the statement that left-handed physics is the same as right-handed physics. Spatial parity is that a left-handed engine works in exactly the same way as a right-handed engine. Violation of spatial parity means that left-handed physics does not work in the same way as right-handed physics. Since handedness is the arbitrary imposition of a particular co-ordinate system in our 4-dimensional space-time, we might expect that spatial parity could not be violated in physics. The existence of left-handed neutrinos and the non-existence of right-handed neutrinos is an apparent violation of spatial parity[66]. Right-handed physics has no neutrinos; left-handed physics has neutrinos; left-hand physics is not the same as right-hand physics. Spatial parity is violated in the universe.

---

[66] There is a lot less anti-matter than matter in the universe. There are a lot less anti-neutrinos than neutrinos in the universe. Thus, the fact that anti-neutrinos are all right-handed does not rebalance the universe.

That spatial parity is violated in the universe does not mean that spatial parity is violated within our 4-dimensional space-time unless our 4-dimensional space-time is the only type of space in the universe. We are of the view that neutrinos exist outside of 4-dimensional space-time[67]. We are of the view that spatial parity cannot be violated within 4-dimensional space-time. Allowing that neutrinos exist outside of 4-dimensional space-time avoids the impossible contradiction of the violation of spatial parity within a space in which spatial parity cannot be violated.

Charge-conjugation and violation of charge-conjugation:

Charge-conjugation is another kind of chirality. We can 'reflect' physical processes done with matter into the same physical processes done with anti-matter.

Charge conjugation is the reversal of all quantum numbers associated with a system. Under charge conjugation, the sign of the electric charge is reversed, the intrinsic parity is reversed, and the direction of the axis of intrinsic angular momentum with respect to the particle's velocity is reversed. For massive particles, which necessarily travel at less than the speed of light, the reversal of the axis of intrinsic angular momentum with respect to velocity is meaningless because, for particles moving at less than the speed of light, two observers moving at different velocities need not necessarily agree on the direction of the velocity of the particle.

We can construct an anti-hydrogen[68] atom from an anti-proton and a positron just as we can construct a matter hydrogen atom from a proton and an electron[69].[70] Such 'reflection' from matter to anti-matter, or from anti-matter to matter, is called charge-conjugation. We believe that an anti-proton is an exact charge conjugation reflection of a proton. In 2017, the Baryon Anti-baryon Symmetry Experiment, BASE, collaboration at CERN reported measurements of the magnetic g-factors (roughly, gyromagnetic ratios) of the proton, 2.792847350(9), and the anti-proton, 2.7928465(23), to be the same within experimental error to very high precision[71]. This is powerful evidence that an anti-proton is an exact charge conjugate reflection of a proton – but see the next section.

It might be a surprise to discover that there are massive particles, like neutral kaons, which are both a kaon and an anti-kaon at the same time and that there is violation of charge-conjugation in the universe. More simply, the imbalance between the amount of matter and the amount of anti-matter in the universe is a violation of the 'matter anti-matter reflection'; it is a violation of charge-conjugation. There are other observed violations of charge-conjugation.

Aside:

The term charge-conjugation came from the idea that electrically charged fermions like the electron are represented by a pair of complex numbers, $\mathbb{C}^2$, and that the corresponding anti-particle is represented by

---

[67] Nothing can travel at the speed of light inside 4-dimensional space-time.

[68] See: Nature 26th Jan 2017 Vol 541 No. 7638.

[69] See: Nature December 2016 reported in CERN Courier Vol 57 No 1 Jan/Feb 2017.

[70] The spectral lines of anti-hydrogen match the spectral lines of matter hydrogen to 2×10^{-10} thereby verifying CPT invariance to this precision : CERN Courier Vol 57 No 1 Jan/Feb 2017.

[71] Reported in CERN Courier Vol 57 No 2 March 2017.

a pair of conjugate complex numbers. Since electrical charge is a real number, we take a different view of electrical charge from the 'imaginary bits of a pair of complex numbers' view, but that's for later.

Charge-conjugation parity invariance:

Charge-conjugation parity invariance (spatial parity), CP invariance, is another kind of chirality. If the universe was CP invariant, then left-handed physics done with matter would be exactly the same as right-handed physics done with anti-matter. CP invariance involves both anti-matter and spatial parity.

The reader might have heard that K-mesons (also known as kaons) and B-mesons do not respect CP invariance. The violation of CP invariance by kaons and B-mesons means that left-handed physics done with matter is not the same as right-handed physics done with anti-matter. Since mesons are pairs of quarks, we say that CP violation has been observed in the quark sector of the Standard Model.

In 2017, the LHCb experiment at CERN reported tantalising evidence that baryons, which are triples of quarks, behave differently from anti-baryons thereby violating CP Invariance[72]. Before this, violation of CP invariance had been seen in only mesons which are pairs of quarks.

CP violation has not yet been observed to be associated with neutrinos or electrons. There are experiments currently looking for CP violation in the electron and neutrino sector of the Standard Model, but it has not yet been observed. We say that CP violation has not been observed in the lepton sector of the Standard Model. We have right-handed anti-neutrinos and left-handed neutrinos; this is exactly what we should have if neutrinos respect CP invariance.

If the universe was CP invariant, there would be as much right-handed anti-matter in the universe as there is left-handed matter in the universe; this is not the case; the universe is primarily matter with only a sprinkling of anti-matter. Violation of CP invariance has been posited as an explanation of the matter anti-matter imbalance of the universe but measurements of the amount of violation of CP invariance indicate that there is not enough violation of CP invariance to explain the amount of the matter/anti-matter imbalance in the universe.

In general:

Violation of parity means that there is an imbalance between left-hand physics and right-hand physics in the universe.

Violation of charge-conjugation means that there is an imbalance between matter physics and anti-matter physics in the universe.

Violation of CP invariance means that there is an imbalance between left-handed matter physics and right-handed anti-matter physics in the universe.

---

[72] Reported in CERN Courier Vol 57 No 2 March 2017.

The Quaternion E-field and the Quaternion B-field

We will take a look at the 4-dimensional $C_2 \times C_2$ algebras, the 4-dimensional Clifford algebras, which are the quaternions $\mathbb{H}_{L\chi}$ & $\mathbb{H}_{R\chi}$.

The U(1) gauge theory:

Traditionally, the locally varying phase of the complex plane, the $U(1)$, phase, is associated with the existence of the electromagnetic photon. There is no chirality in the complex numbers, $\mathbb{C}$.

The SU(2) gauge theory:

Traditionally, the locally varying phase of the quaternions, the $SU(2)$ phase, is associated with the existence of the three weak force bosons, W^+, W^-, Z^0.

Quaternions in particle physics:

We remind the reader that, there are two types of quaternions which are the left-chiral quaternions and the right-chiral quaternions. The chirality of a quaternion is the 'handedness' of the commutation relations, either $SU(2)_{L\chi}$ or $SU(2)_{R\chi}$. In contrast to the conventional particle physics view of intrinsic spin, quaternion chirality does not refer to an axis.

Left-chiral quaternions:

The left-chiral quaternion differential operator is:

$$d_{(L\chi)} = \begin{bmatrix} \partial_t & -\partial_x & -\partial_y & -\partial_z \\ \partial_x & \partial_t & \partial_z & -\partial_y \\ \partial_y & -\partial_z & \partial_t & \partial_x \\ \partial_z & \partial_y & -\partial_x & \partial_t \end{bmatrix} \tag{10.1}$$

Left-chiral quaternion potential:

We have some choice over the signs we assign to elements of the left-chiral quaternion potential. The choice is arbitrary, but we must respect the distribution of minus signs within the matrix if we are to keep the left-chiral algebraic structure.

By Noether's theorem, there is a conserved charge and a conserved current associated with rotation in every division algebra space acting as a gauge space over space-time. The conserved charge is a real number, and the conserved current is an ordered triplet of three numbers. In the case of the two quaternion algebras, both left-chiral and right-chiral, we take this conserved charge to be electric charge and we take the conserved current to be intrinsic angular momentum.

Since we take the view that the real variable in a quaternion potential, ϕ, is electric charge, we will take this to be the electric charge of the electron in due course.

We choose the left-chiral quaternion matter potential to be:

$$\Phi_{L\chi}^{Matter} = \begin{bmatrix} \phi & -A_x & -A_y & -A_z \\ A_x & \phi & A_z & -A_y \\ A_y & -A_z & \phi & A_x \\ A_z & A_y & -A_x & \phi \end{bmatrix} \tag{10.2}$$

The reader might think we have cheated by using the conjugate quaternion for the potential. This left-chiral quaternion matter potential, (10.2), leads to an E-field which matches the conventionally, if arbitrarily, defined classical electric field. We could have used a different potential, but we might as well stick with the conventional definition of the classical electric field.

We non-commutatively differentiate the left-chiral matter potential to give the E-field and the B-field. The E-field of a left-chiral quaternion potential is a left-chiral quaternion matrix. The E-field of the left-chiral quaternion matter potential, (10.2), is:

$$E_{[1,1]} = \frac{\partial \phi}{\partial t} - \frac{\partial A_x}{\partial x} - \frac{\partial A_y}{\partial y} - \frac{\partial A_z}{\partial z} \tag{10.3}$$

$$E_{[1,2]} = -\frac{\partial \phi}{\partial x} - \frac{\partial A_x}{\partial t}$$

$$E_{[1,3]} = -\frac{\partial \phi}{\partial y} - \frac{\partial A_y}{\partial t} \tag{10.4}$$

$$E_{[1,4]} = -\frac{\partial \phi}{\partial z} - \frac{\partial A_z}{\partial t}$$

$$E_{\mathbb{H}_{L\chi}}^{Matter} = \begin{bmatrix} E_{[1,1]} & E_{[1,2]} & E_{[1,3]} & E_{[1,4]} \\ -E_{[1,2]} & E_{[1,1]} & -E_{[1,4]} & E_{[1,3]} \\ -E_{[1,3]} & E_{[1,4]} & E_{[1,1]} & -E_{[1,2]} \\ -E_{[1,4]} & -E_{[1,3]} & E_{[1,2]} & E_{[1,1]} \end{bmatrix} \tag{10.5}$$

The three imaginary elements of the E-field are equal to the arbitrarily defined classical electric field. We draw the reader's attention to the fact that the classical electric field is a space-time curl with both signs being the same; of course, space-time curls are linear acceleration forces.

The corresponding B-field of the left-chiral quaternion matter potential, (10.2), is:

$$B_{[1,1]} = 0$$

$$B_{[1,2]} = \frac{\partial A_z}{\partial y} - \frac{\partial A_y}{\partial z}$$

$$B_{[1,3]} = \frac{\partial A_x}{\partial z} - \frac{\partial A_z}{\partial x} \qquad (10.6)$$

$$B_{[1,4]} = \frac{\partial A_y}{\partial x} - \frac{\partial A_x}{\partial y}$$

$$B_{\mathbb{H}_{L\chi}}^{Matter} = \begin{bmatrix} 0 & B_{[1,2]} & B_{[1,3]} & B_{[1,4]} \\ -B_{[1,2]} & 0 & -B_{[1,4]} & B_{[1,3]} \\ -B_{[1,3]} & B_{[1,4]} & 0 & -B_{[1,2]} \\ -B_{[1,4]} & -B_{[1,3]} & B_{[1,2]} & 0 \end{bmatrix} \qquad (10.7)$$

The three imaginary elements of the B-field are equal to the conventionally, if arbitrarily, defined classical magnetic field. Note that the B-field is a Euclidean spatial curl. We will later present the view that the neutrino field is to the electron field as the classical magnetic field is to the classical electric field.

Left-chiral anti-matter:
The electric charge, which is a real number, of a positron is opposite to the electric charge of an electron. We take the real quaternion variable of a positron to be the negative of the electron. If we are taking the electric field of the electron to be of the form given as the three imaginary parts of the E-field, (10.4), then the electric field of the positron must be of the same form but of opposite sign.

An important point:
Because the quaternions are a division algebra when the real variable is negative as well as when the real variable is positive, we can form a potential with a negative real variable. We cannot do this with the A_3 algebras nor with the 8-dimensional Clifford algebras.

Aside:
We could have chosen a potential with the signs of the imaginary elements to be reversed; this would have given a different form of electric field. Had we done this, we would have involved the time variable is a purely spatial type of curl. Since time cannot be reversed, we would be in a mess. The time variable must be involved in a space-time curl. The nature of a space-time curl can be found by differentiating the hyperbolic complex numbers, $\mathbb{S}$; it is two plus signs as opposed to a Euclidean space curl which is a plus sign and a minus sign as is found by differentiating the Euclidean complex numbers, $\mathbb{C}$.

The left-chiral quaternion anti-matter potential:

We choose the left-chiral quaternion anti-matter potential to be:

$$\Phi_{L\chi}^{Anti-Matter} = \begin{bmatrix} -\phi & A_x & A_y & A_z \\ -A_x & -\phi & -A_z & A_y \\ -A_y & A_z & -\phi & -A_x \\ -A_z & -A_y & A_x & -\phi \end{bmatrix} \tag{10.8}$$

We have reversed the sign of the charge, the real variable, and we have conjugated, reversed the sign of, the imaginary variables which we associate with the intrinsic angular momentum.

We have taken anti-matter to be the reverse of the sign of electric charge and the reverse of the direction of the axis of intrinsic angular momentum with respect to velocity.

The E-field of the left-chiral quaternion anti-matter potential, (10.8), is:

$$E_{[1,1]} = -\frac{\partial \phi}{\partial t} + \frac{\partial A_x}{\partial x} + \frac{\partial A_y}{\partial y} + \frac{\partial A_z}{\partial z} \tag{10.9}$$

$$E_{[1,2]} = \frac{\partial \phi}{\partial x} + \frac{\partial A_x}{\partial t}$$

$$E_{[1,3]} = \frac{\partial \phi}{\partial y} + \frac{\partial A_y}{\partial t} \tag{10.10}$$

$$E_{[1,4]} = \frac{\partial \phi}{\partial z} + \frac{\partial A_z}{\partial t}$$

$$E_{\mathbb{H}_{L\chi}}^{Anti-Matter} = \begin{bmatrix} E_{[1,1]} & E_{[1,2]} & E_{[1,3]} & E_{[1,4]} \\ -E_{[1,2]} & E_{[1,1]} & -E_{[1,4]} & E_{[1,3]} \\ -E_{[1,3]} & E_{[1,4]} & E_{[1,1]} & -E_{[1,2]} \\ -E_{[1,4]} & -E_{[1,3]} & E_{[1,2]} & E_{[1,1]} \end{bmatrix} \tag{10.11}$$

The three imaginary elements of this E-field, (10.10), are opposite to the matter E-field, (10.4); this fits with the reversal of the electric field.

The corresponding B-field of the left-chiral quaternion anti-matter potential, (10.8), is:

$$B_{[1,1]} = 0$$

$$B_{[1,2]} = -\frac{\partial A_z}{\partial y} + \frac{\partial A_y}{\partial z}$$

$$B_{[1,3]} = -\frac{\partial A_x}{\partial z} + \frac{\partial A_z}{\partial x} \tag{10.12}$$

$$B_{[1,4]} = -\frac{\partial A_y}{\partial x} + \frac{\partial A_x}{\partial y}$$

$$B_{\mathbb{H}_{L\chi}}^{Anti-Matter} = \begin{bmatrix} 0 & B_{[1,2]} & B_{[1,3]} & B_{[1,4]} \\ -B_{[1,2]} & 0 & -B_{[1,4]} & B_{[1,3]} \\ -B_{[1,3]} & B_{[1,4]} & 0 & -B_{[1,2]} \\ -B_{[1,4]} & -B_{[1,3]} & B_{[1,2]} & 0 \end{bmatrix} \qquad (10.13)$$

This is the opposite of the left-chiral matter B-field, (10.6).

The right-chiral quaternions:

The right-chiral quaternion differential operator is (distribution of minus signs):

$$d_{(R\chi)} = \begin{bmatrix} \partial_t & -\partial_x & -\partial_y & -\partial_z \\ \partial_x & \partial_t & -\partial_z & \partial_y \\ \partial_y & \partial_z & \partial_t & -\partial_x \\ \partial_z & -\partial_y & \partial_x & \partial_t \end{bmatrix} \qquad (10.14)$$

We choose the right-chiral quaternion matter potential to be:

$$\Phi_{R\chi}^{Matter} = \begin{bmatrix} \phi & -A_x & -A_y & -A_z \\ A_x & \phi & -A_z & A_y \\ A_y & A_z & \phi & -A_x \\ A_z & -A_y & A_x & \phi \end{bmatrix} \qquad (10.15)$$

This, (10.14) &(10.15), gives the E-field of the right-chiral quaternion matter potential with components:

$$E_{[1,1]} = \frac{\partial\phi}{\partial t} - \frac{\partial A_x}{\partial x} - \frac{\partial A_y}{\partial y} - \frac{\partial A_z}{\partial z} \qquad (10.16)$$

$$E_{[1,2]} = -\frac{\partial\phi}{\partial x} - \frac{\partial A_x}{\partial t}$$

$$E_{[1,3]} = -\frac{\partial\phi}{\partial y} - \frac{\partial A_y}{\partial t} \qquad (10.17)$$

$$E_{[1,4]} = -\frac{\partial\phi}{\partial z} - \frac{\partial A_z}{\partial t}$$

$$E_{\mathbb{H}_{R\chi}}^{Matter} = \begin{bmatrix} E_{[1,1]} & E_{[1,2]} & E_{[1,3]} & E_{[1,4]} \\ -E_{[1,2]} & E_{[1,1]} & E_{[1,4]} & -E_{[1,3]} \\ -E_{[1,3]} & -E_{[1,4]} & E_{[1,1]} & E_{[1,2]} \\ -E_{[1,4]} & E_{[1,3]} & -E_{[1,2]} & E_{[1,1]} \end{bmatrix} \qquad (10.18)$$

This is identical to the E-field of the left-chiral quaternion matter potential, (10.3) & (10.4), except for the chirality which is as expressed in the distribution of the minus signs within the matrix.

This, (10.14) & (10.15), gives the B-field of the right-chiral quaternion matter potential as:

$$B_{[1,1]} = 0$$

$$B_{[1,2]} = -\frac{\partial A_z}{\partial y} + \frac{\partial A_y}{\partial z}$$

$$B_{[1,3]} = -\frac{\partial A_x}{\partial z} + \frac{\partial A_z}{\partial x} \tag{10.19}$$

$$B_{[1,4]} = -\frac{\partial A_y}{\partial x} + \frac{\partial A_x}{\partial y}$$

$$B_{\mathbb{H}_{R\chi}}^{Matter} = \begin{bmatrix} 0 & B_{[1,2]} & B_{[1,3]} & B_{[1,4]} \\ -B_{[1,2]} & 0 & B_{[1,4]} & -B_{[1,3]} \\ -B_{[1,3]} & -B_{[1,4]} & 0 & B_{[1,2]} \\ -B_{[1,4]} & B_{[1,3]} & -B_{[1,2]} & 0 \end{bmatrix} \tag{10.20}$$

This is the opposite of the B-field of the left-chiral quaternion matter potential, (10.6).

Right-chiral anti-matter:

We choose the right-chiral quaternion anti-matter potential to be:

$$\Phi_{R\chi}^{Anti-Matter} = \begin{bmatrix} -\phi & A_x & A_y & A_z \\ -A_x & -\phi & A_z & -A_y \\ -A_y & -A_z & -\phi & A_x \\ -A_z & A_y & -A_x & -\phi \end{bmatrix} \tag{10.21}$$

The E-field of the right-chiral quaternion anti-matter potential, (10.21), is:

$$E_{[1,1]} = -\frac{\partial \phi}{\partial t} + \frac{\partial A_x}{\partial x} + \frac{\partial A_y}{\partial y} + \frac{\partial A_z}{\partial z}$$

$$E_{[1,2]} = \frac{\partial \phi}{\partial x} + \frac{\partial A_x}{\partial t}$$

$$E_{[1,3]} = \frac{\partial \phi}{\partial y} + \frac{\partial A_y}{\partial t} \tag{10.22}$$

$$E_{[1,4]} = \frac{\partial \phi}{\partial z} + \frac{\partial A_z}{\partial t}$$

$$E_{\mathbb{H}_{R\chi}}^{Anti-Matter} = \begin{bmatrix} E_{[1,1]} & E_{[1,2]} & E_{[1,3]} & E_{[1,4]} \\ -E_{[1,2]} & E_{[1,1]} & E_{[1,4]} & -E_{[1,3]} \\ -E_{[1,3]} & -E_{[1,4]} & E_{[1,1]} & E_{[1,2]} \\ -E_{[1,4]} & E_{[1,3]} & -E_{[1,2]} & E_{[1,1]} \end{bmatrix} \tag{10.23}$$

This, (10.22), is identical to the E-field of the left-chiral quaternion anti-matter potential above, (10.9) & (10.10).

The B-field of the right-chiral quaternion anti-matter potential, (10.21), is:

$$B_{[1,1]} = 0$$

$$B_{[1,2]} = \frac{\partial A_z}{\partial y} - \frac{\partial A_y}{\partial z}$$

$$B_{[1,3]} = \frac{\partial A_x}{\partial z} - \frac{\partial A_z}{\partial x} \qquad (10.24)$$

$$B_{[1,4]} = \frac{\partial A_y}{\partial x} - \frac{\partial A_x}{\partial y}$$

$$B_{\mathbb{H}_{R\chi}}^{Anti-Matter} = \begin{bmatrix} 0 & B_{[1,2]} & B_{[1,3]} & B_{[1,4]} \\ -B_{[1,2]} & 0 & B_{[1,4]} & -B_{[1,3]} \\ -B_{[1,3]} & -B_{[1,4]} & 0 & B_{[1,2]} \\ -B_{[1,4]} & B_{[1,3]} & -B_{[1,2]} & 0 \end{bmatrix} \qquad (10.25)$$

This, (10.24), is identical to the left-chiral quaternion matter B-field (10.6).

Electrons and neutrinos:

We take the E-fields to be electron fields. We take the B-fields to be neutrino fields. At this stage, we thus have left-chiral electrons and right-chiral electrons and left-chiral positrons and right-chiral positrons, and we thus have left-chiral neutrinos and right-chiral neutrinos and left-chiral anti-neutrinos and right-chiral anti-neutrinos. Things will change when we take the superimposition of these fields, which we do by adding them – next chapter.

The discovery of CP invariance:

Bringing this all together, (10.3) & (10.4) & (10.9) & (10.10) & (10.16) & (10.17) & (10.22), we have:

$$E_{L\chi}^{Matter} = E_{R\chi}^{Matter} = -E_{L\chi}^{Anti-Matter} = -E_{R\chi}^{Anti-Matter} \qquad (10.26)$$

We see that the E-field changes with swapping the sign of the matter electric charge for the anti-matter electric charge. We see that the E-field is 'immune' from change of chirality. The matter/anti-matter nature of these E-fields, electrons, is a simple change of sign of the electric charge and a change of sign of the intrinsic angular momentum.

We also have, (10.6) & (10.12) & (10.19) & (10.24):

$$B_{L\chi}^{Matter} = -B_{R\chi}^{Matter} = -B_{L\chi}^{Anti-Matter} = B_{R\chi}^{Anti-Matter} \qquad (10.27)$$

The equation $B_{L\chi}^{Matter} = B_{R\chi}^{Antimatter}$ means that the left-handed matter B-field is the same as the right-handed anti-matter B-field – this is CP invariance.

The equation $B_{L\chi}^{Matter} = -B_{R\chi}^{Matter}$ means that the left-handed matter B-field is not the same as the right-handed matter B-field – this is violation of spatial parity.

CP invariance:

From (10.27), we extract:

$$B_{L\chi}^{Matter} = B_{R\chi}^{Anti-Matter}$$
$$B_{R\chi}^{Matter} = B_{L\chi}^{Anti-Matter}$$

(10.28)

We see that a left-chiral matter B-field, neutrino, is equal to a right-chiral anti-matter B-field, neutrino. We see that the anti-matter aspect of neutrinos is a combination of chirality and change of sign of the electric charge. This concurs with the violation of parity and respect for charge-parity which are observed properties of the neutrino.

The anti-particle of the electron is simply the positron. Great, we all understand that; the positron is a 'reflection' of the electron in terms of electric charge and intrinsic angular momentum. Neutrinos have no electric charge.

We see from (10.28) that we cannot apply the same kind of matter & anti-matter reflection based on only electric charge and intrinsic angular momentum to the neutrino – neutrinos are without electric charge. There are still two types of neutrino – they come from four different potentials, but neutrinos are not simply matter & anti-matter reflections of each other based on electric charge and intrinsic angular momentum; they are chirality reflections of each other.

Two types of chirality:

We formed the anti-matter potentials by reversing the sign of the electric charge, the real variable, and by reversing the signs of the intrinsic angular momentum, the three imaginary variables. This produced two types of 'reflection' in quaternion space.

We can change an electron into a positron by only one route; we do this by changing the sign of the electric charge and changing the sign of the intrinsic angular momentum.

We can change a neutrino into an anti-neutrino by two separate roots; either we can change the sign of the electric charge and change the sign of the intrinsic angular momentum or we can change the chirality. Of course, neutrinos have no electric charge, and so we can ignore this part of the change.

We have two different ways of defining anti-matter – the electron way or the neutrino way.

Breaking Algebras

Each type of division algebra space, spinor space, has its own type of rotational symmetry. We will break the algebras, and, in so doing, we will necessarily break the symmetries of those algebras. One cannot break a symmetry without also breaking the algebra because breaking a symmetry is breaking the division algebra space, the spinor space, which holds that symmetry and vice-versa.

We break the algebras by taking a superimposition of them. This is just adding all the different types of the same algebra. For example, the quaternion potential will be broken by adding the left-chiral quaternion potential (10.2) and the right-chiral quaternion potential (10.15).

We cannot add two elements of different algebras even if the two algebras are algebraically isomorphic to each other because addition, as multiplication, does not properly exist outside of an algebraic structure, division algebra[73]. When we add the two quaternion potentials, we are adding the real numbers which are the elements of the matrices; in doing this, we lose the matrix structure; we 'break the algebra'; by this we mean that the result is not an element of any algebraic matrix form. It seems that we must 'break the algebras' for them to appear in our $\mathbb{R}^4$ 4-dimensional space-time.

When we form the sum of two or more copies of the same kind of algebra, we call the result the 'emergent space' of that algebra. For example, the emergent quaternion matter potential is:

$$\Phi_{L\chi}^{Matter} + \Phi_{R\chi}^{Matter} = \begin{bmatrix} \phi & -A_x & -A_y & -A_z \\ A_x & \phi & A_z & -A_y \\ A_y & -A_z & \phi & A_x \\ A_z & A_y & -A_x & \phi \end{bmatrix} + \begin{bmatrix} \phi & -A_x & -A_y & -A_z \\ A_x & \phi & -A_z & A_y \\ A_y & A_z & \phi & -A_x \\ A_z & -A_y & A_x & \phi \end{bmatrix} = 2\begin{bmatrix} \phi & -A_x & -A_y & -A_z \\ A_x & \phi & 0 & 0 \\ A_y & 0 & \phi & 0 \\ A_z & 0 & 0 & \phi \end{bmatrix}$$

(11.1)

If we like, we can neglect the 2 because it is no more than a scaling factor. We take the ϕ of (11.1) to be the electric charge of the electron. The A_i variables taken separately, are generator matrices for three 2-dimensional rotations.

$$\exp\left(\begin{bmatrix} 0 & 0 & A_y & 0 \\ 0 & 0 & 0 & 0 \\ A_y & 0 & 0 & 0 \\ 0 & 0 & 0 & 0 \end{bmatrix}\right) = \begin{bmatrix} \cos A_y & 0 & -\sin A_y & 0 \\ 0 & 1 & 0 & 0 \\ \sin A_y & 0 & \cos A_y & 0 \\ 0 & 0 & 0 & 1 \end{bmatrix} \qquad (11.2)$$

---

[73] The form of the two different matrices is not conserved by addition – no additive closure.

Aside:

If we add the quaternion matter potentials to the quaternion anti-matter potentials, we get a zero potential. It is as if electrons and positrons mutually annihilate.

The nature of emergent spaces:

Imaginary variables exist within only division algebras. An emergent space is not a division algebra and therefore has no imaginary variables, but the original variables are still there; they must be real variables. The variables in an emergent space are all real. Instead of a division algebra space like quaternion space, a 'broken algebra' space, emergent space, is an ordered n-tuple of real numbers. In the case of the 4-dimensional emergent quaternion space, the space is $\mathbb{R}^4$.

Each of the division algebras which are added to form the emergent space contains a single n-dimensional angle. The emergent space gets one angle from each of the algebras which were added to form the emergent space. In the quaternion case, the 4-dimensional emergent quaternion space has only two angles[74].

The distance function of an emergent space is simply the sum of the individual distance functions of the algebras which were added to form the emergent space. This is just a sum of real numbers. In the quaternion case, we have:

$$d^2 = t^2 + x^2 + y^2 + z^2$$
$$d^2 = t^2 + x^2 + y^2 + z^2 \quad +$$
$$- - - - - - - - - -$$
$$2d^2 = 2t^2 + 2x^2 + 2y^2 + 2z^2$$
$$d^2 = t^2 + x^2 + y^2 + z^2$$

(11.3)

The form of the distance function of the emergent quaternion space is identical to the distance function of the quaternion space; this distance function allows 4-dimensional quaternion rotation, and so it seems that the emergent quaternion space has two 4-dimensional angles. Of course, quaternion rotation is double cover rotation. We opine that this is why the gyromagnetic ratio of the electron is two[75].

Superimpositions:

We are going to form the superimpositions of the B-fields and the E-fields. In modern particle physics parlance, these would be called super-positions. In modern particle physics, these super-positions might be written as:

---

[74] This is conceptually quite shocking. We have four axes which form six pairs of axes, but we have only two angles to fit into these six pairs of axes. This might be what leads to the directional quantitisation of electron spin, but we opine that electron spin is chirality.

[75] The gyromagnetic ratio of the electron is not exactly 2. The difference is to do with quantum fluctuations of the vacuum and is well understood.

$$e^- = \left| \mathbb{H}_{L\chi}^{Matter} \right\rangle_{E-field} + \left| \mathbb{H}_{R\chi}^{Matter} \right\rangle_{E-field} \qquad e^+ = \left| \mathbb{H}_{L\chi}^{Anti-Matter} \right\rangle_{E-field} + \left| \mathbb{H}_{R\chi}^{Anti-Matter} \right\rangle_{E-field}$$

$$\nu = \left| \mathbb{H}_{L\chi}^{Matter} \right\rangle_{B-field} + \left| \mathbb{H}_{R\chi}^{Matter} \right\rangle_{B-field} \qquad \bar{\nu} = \left| \mathbb{H}_{L\chi}^{Anti-Matter} \right\rangle_{B-field} + \left| \mathbb{H}_{R\chi}^{Anti-Matter} \right\rangle_{B-field} \tag{11.4}$$

The emergent quaternion differentials:

We have, by adding (10.4) and (10.17), the emergent matter E-field:

$$E_{L\chi}^{Matter} + E_{R\chi}^{Matter} =$$

$$E_{Emergent}^{Matter} = 2 \begin{bmatrix} Div & -\dfrac{\partial \phi}{\partial x} - \dfrac{\partial A_x}{\partial t} & -\dfrac{\partial \phi}{\partial y} - \dfrac{\partial A_y}{\partial t} & -\dfrac{\partial \phi}{\partial z} - \dfrac{\partial A_z}{\partial t} \\[2mm] \dfrac{\partial \phi}{\partial x} + \dfrac{\partial A_x}{\partial t} & Div & 0 & 0 \\[2mm] \dfrac{\partial \phi}{\partial y} + \dfrac{\partial A_y}{\partial t} & 0 & Div & 0 \\[2mm] \dfrac{\partial \phi}{\partial z} + \dfrac{\partial A_z}{\partial t} & 0 & 0 & Div \end{bmatrix} \tag{11.5}$$

$$Div = \frac{\partial \phi}{\partial t} - \frac{\partial A_x}{\partial x} - \frac{\partial A_y}{\partial y} - \frac{\partial A_z}{\partial z}$$

We take this to be the electron field.

We have, by adding (10.10) and (10.22), the emergent anti-matter E-field:

$$E_{L\chi}^{Anti-Matter} + E_{R\chi}^{Anti-Matter} =$$

$$E_{Emergent}^{Anti-Matter} = 2 \begin{bmatrix} Div & \dfrac{\partial \phi}{\partial x} + \dfrac{\partial A_x}{\partial t} & \dfrac{\partial \phi}{\partial y} + \dfrac{\partial A_y}{\partial t} & \dfrac{\partial \phi}{\partial z} + \dfrac{\partial A_z}{\partial t} \\[2mm] -\dfrac{\partial \phi}{\partial x} - \dfrac{\partial A_x}{\partial t} & Div & 0 & 0 \\[2mm] -\dfrac{\partial \phi}{\partial y} - \dfrac{\partial A_y}{\partial t} & 0 & Div & 0 \\[2mm] -\dfrac{\partial \phi}{\partial z} - \dfrac{\partial A_z}{\partial t} & 0 & 0 & Div \end{bmatrix} \tag{11.6}$$

$$Div = -\frac{\partial \phi}{\partial t} + \frac{\partial A_x}{\partial x} + \frac{\partial A_y}{\partial y} + \frac{\partial A_z}{\partial z}$$

We see that the electric field is reversed in (11.6) compared to (11.5).

We note that both of the emergent E-fields are not chiral. The chiral part of the quaternion matrix is in the bottom right-hand 3×3 corner of a quaternion matrix. We see that these are zeros in the emergent E-fields.

We have, by adding (10.6) and (10.19), the emergent matter B-field:

$$B_{L\chi}^{Matter} + B_{R\chi}^{Matter} =$$

$$B_{Emergent}^{Matter} = \begin{bmatrix} 0 & 0 & 0 & 0 \\ 0 & 0 & -\dfrac{\partial A_y}{\partial x} + \dfrac{\partial A_x}{\partial y} & \dfrac{\partial A_x}{\partial z} - \dfrac{\partial A_z}{\partial x} \\ 0 & \dfrac{\partial A_y}{\partial x} - \dfrac{\partial A_x}{\partial y} & 0 & -\dfrac{\partial A_z}{\partial y} + \dfrac{\partial A_y}{\partial z} \\ 0 & -\dfrac{\partial A_x}{\partial z} + \dfrac{\partial A_z}{\partial x} & \dfrac{\partial A_z}{\partial y} - \dfrac{\partial A_y}{\partial z} & 0 \end{bmatrix} \qquad (11.7)$$

This, (11.7), is a field which has zero electric charge – zero real variable. This field, (11.7), is a left-chiral field. We take this to be the neutrino field.

We have, by adding (10.12) and (10.24), the emergent anti-matter B-field:

$$B_{L\chi}^{Anti-Matter} + B_{R\chi}^{Anti-Matter} =$$

$$B_{Emergent}^{Anti-Matter} = \begin{bmatrix} 0 & 0 & 0 & 0 \\ 0 & 0 & \dfrac{\partial A_y}{\partial x} - \dfrac{\partial A_x}{\partial y} & -\dfrac{\partial A_x}{\partial z} - \dfrac{\partial A_z}{\partial x} \\ 0 & -\dfrac{\partial A_y}{\partial x} + \dfrac{\partial A_x}{\partial y} & 0 & \dfrac{\partial A_z}{\partial y} - \dfrac{\partial A_y}{\partial z} \\ 0 & \dfrac{\partial A_x}{\partial z} - \dfrac{\partial A_z}{\partial x} & -\dfrac{\partial A_z}{\partial y} + \dfrac{\partial A_y}{\partial z} & 0 \end{bmatrix} \qquad (11.8)$$

This, (11.7), is a field which has no electric charge. This field, (11.7), is a right-chiral field. We take this to be the anti-neutrino field – remember 'anti' in terms of chirality not in terms of electric charge.

The leading diagonal of the quaternion algebras, the real variable, coincides with the leading diagonal of the A_3 algebras. We associate the real A_3 variable with both mass (space-time rotation) and classical electric charge (Euclidean rotation), and we associate the real quaternion variable with electric charge. Waving our arms around and talking of coinciding variables, we assume the real quaternion variable manifests itself in our 4-dimensional space-time as mass as well as electric charge. Thus, the neutrino field is massless and electrically neutral. This concurs with observations that the neutrino moves at the speed of light. We note that the neutrino field squared (multiplied by its own conjugate) has a non-zero real part – it is massive. Thus neutrinos can oscillate between generations.

The Dirac equation:
We have covered this in more detail elsewhere[76].

Dirac sought a solution to the relativistic energy momentum relation:

---

[76] See: Dennis Morris : The Quaternion Dirac Equation.

$$E^2 = p_x^2 + p_y^2 + p_z^2 + m^2 \tag{11.9}$$

He did this by seeking a square root of the RHS of (11.9) within one of the five 16-dimensional Clifford algebras. This led him to the famous Dirac equation in which four elements of the 16-dimensional Clifford algebra, one basis vector and three basis bi-vectors, are represented by the four famous gamma matrices. The bi-linear covariants associated with the Dirac equation are then the other basis elements of the 16-dimensional Clifford algebra.

A simpler way to solve the relativistic energy momentum relation is to see that the RHS of (11.9) is a quaternion inner product (norm). There are four solutions:

$$\Phi_{R\chi}\Phi_{R\chi}^* = \begin{bmatrix} \phi & -p_x & -p_y & -p_z \\ p_x & \phi & -p_z & p_y \\ p_y & p_z & \phi & -p_x \\ p_z & -p_y & p_x & \phi \end{bmatrix} \begin{bmatrix} \phi & p_x & p_y & p_z \\ -p_x & \phi & p_z & -p_y \\ -p_y & -p_z & \phi & p_x \\ -p_z & p_y & -p_x & \phi \end{bmatrix} \tag{11.10}$$

$$\Phi_{L\chi}^{Anti}\Phi_{L\chi}^{Anti*} \quad \& \quad \Phi_{R\chi}^{Anti}\Phi_{R\chi}^{Anti*} \quad \& \quad \Phi_{L\chi}\Phi_{L\chi}^* \tag{11.11}$$

Who needs 16-dimensional Clifford algebras to solve (11.9)?

Solving the relativistic energy momentum relation using quaternions presents an explanation of the neutrino mass problem and the handedness of the neutrino.

Aside: Emergent spaces formed by adding equal chirality:
We have:

$$E_{L\chi}^{Matter} + E_{L\chi}^{Anti-Matter} =$$

$$E_{L\chi}^{Emergent} = 2 \begin{bmatrix} 0 & 0 & 0 & 0 \\ 0 & 0 & \dfrac{\partial\phi}{\partial z} + \dfrac{\partial A_z}{\partial t} & -\dfrac{\partial\phi}{\partial y} - \dfrac{\partial A_y}{\partial t} \\ 0 & -\dfrac{\partial\phi}{\partial z} - \dfrac{\partial A_z}{\partial t} & 0 & \dfrac{\partial\phi}{\partial x} + \dfrac{\partial A_x}{\partial t} \\ 0 & \dfrac{\partial\phi}{\partial y} + \dfrac{\partial A_y}{\partial t} & -\dfrac{\partial\phi}{\partial x} - \dfrac{\partial A_x}{\partial t} & 0 \end{bmatrix} \tag{11.12}$$

This is a massless right-chiral electric field. We have no idea what to make of this – photon perhaps.

The Emergent Space of the A₃ Algebras

The commutative finite group $C_2 \times C_2$ holds only eight non-commutative division algebras; these are the two quaternion algebras and the six A_3 algebras. There are three pairs of A_3 algebras. Each pair is comprised of one left-chiral algebra and one right-chiral algebra. We give an example of an A_3 algebra:

$$SAS_{L\chi} = \exp\left(\begin{bmatrix} a & b & c & d \\ b & a & d & c \\ -c & d & a & -b \\ d & -c & -b & a \end{bmatrix}\right) \tag{12.1}$$

Note that an A_3 algebra has two symmetric imaginary variables and one anti-symmetric imaginary variable.

The distance function, norm, of this algebra, (12.1), is:

$$dist^2 = a^2 - b^2 + c^2 - d^2 \tag{12.2}$$

The other types of A_3 algebras have similar distance functions with two plus signs and two minus signs.

The emergent space of the A_3 algebras, formed by adding all six of the A_3 algebras, is unique in that from the whole infinitude of finite groups it is the only emergent space which has the same number of angles as it has pairs of axes.

There are only two emergent spaces which have a distance function such that the emergent space can hold rotations of lesser dimensional spaces. These two spaces are:

  1) the quaternion emergent space
  2) the A_3 emergent space

Both of these emergent spaces emerge from the $C_2 \times C_2$ finite group.

Our space-time:

The distance function of the A_3 emergent space, after adjusting the units, is:

$$dist^2 = t^2 - x^2 - y^2 - z^2 \tag{12.3}$$

This is the same as the observed distance function of our 4-dimensional space-time. This emergent distance function, (12.3), can accommodate six 2-dimensional rotations, three Euclidean rotations and three 2-dimensional space-time rotations; it can do this because setting any two of the variables to zero produces the distance function of a 2-dimensional spinor space (division algebra). Together with the six angles from the six A_3 algebras which were superimposed (added) to form this emergent space, we have an exact fit for our observed 4-dimensional space-time.

Note that the emergent distance function (12.3) cannot hold the 4-dimensional rotations of the A_3 algebras because these 4-dimensional rotations hold invariant distance functions of the form of (12.2). We get only the 2-dimensional bits of these 4-dimensional rotations in our 4-dimensional space-time.

By Noether's theorem, all division algebra spaces, spinor spaces, acting as gauge spaces over our 4-dimensional space-time, when reduced to 2-dimensional sub-algebras, have a conserved charge and a conserved current corresponding to each type of rotation ion the space. The A_3 spaces are a little complicated in that they hold two types of rotation. The anti-symmetric variable in the A_3 algebra corresponds to a Euclidean type of rotation, and the symmetric variables of the A_3 algebra correspond to space-time types of rotation. Noether's theorem predicts a conserved charge and a conserved current for each type of rotation. We take the symmetric charge of the A_3 algebras to be mass. We therefore have three mass charges – one for each pair of A_3 algebras[77]. Because mass is a real scalar, we take the real variable within an A_3 algebra, the variable on the leading diagonal, to be mass. The other symmetric variables we take to be the momentum current. We take the anti-symmetric charge to be the classical electric charge and the anti-symmetric current to be the classical magnetic field.

Electric charge and mass:
The nature of the quaternion algebras is such that the real variable of a quaternion can be positive or negative[78]. It follows that the electric charge can be positive or negative.

The nature of the A_3 algebras is such that the real variable of an A_3 algebra can be only positive – it's the exponential that does this – see (12.1). It follows that we will have only positive mass – no negative mass[79]. We have no negative classical electric charge and no negative classical magnetic field.

The curvature of emergent spaces:
Division algebra spaces (spinor spaces) are flat; they have to be flat to maintain the algebraic structure. Division algebra spaces are so flat that they do not even have zero curvature.

When we break an algebra by superimposition, we break the algebraic structure and so we lose the 'not even zero curvature' nature of the division algebra. This means that there is nothing to hold the emergent space flat. The emergent space is not necessarily curved, but it is not compelled to be flat.

---

[77] We opine that this is why there are three generations of particles, but we do not yet properly understand this area of physics.
[78] We do not have to take the exponential of the matrix to form the algebra.
[79] If we allow that mass 'causes' gravity, there will be no negative gravity.

Chapter 13

The E-field and the B-field of the A₃ Algebras

The important result of this chapter is that we discover our 4-dimensional space-time within the A_3 algebras and that we discover spatial parity associated with our 4-dimensional space-time.

Introduction:

In a previous chapter, we dealt with the quaternion E-fields and the quaternion B-fields. There are only two quaternion algebras; there are six A_3 algebras. In this chapter, we will deal with the A_3 E-fields and the A_3 B-fields, exactly as we dealt with the quaternion E-fields and the quaternion B-fields except that we will have to also take account of having three fields of left-chirality and three fields of right-chirality rather than just one field of each chirality. We do this by adding together all the B-fields and adding together all the E-fields and then by separating out the symmetric parts of the sum and the anti-symmetric parts of the sum. Since the quaternions are wholly anti-symmetric, we did not need to consider the separation into symmetric and anti-symmetric parts in the quaternion case.

Noether's theorem gives a conserved charge for each type of rotation in a gauge space. By separating the symmetric parts and the anti-symmetric parts of the sum of the E-fields and of the sum of the B-fields, we are separating the different 'forces' associated with each of the two types of conserved charge.

The E-fields and the B-fields of the SAS algebras:

The differential operator of the A_3 $SAS_{L\chi}$ algebra is formed by adding the inverses of the individual variables and replacing the variable with the partial differential with respect to that variable:

$$d_{L\chi}^{SAS} = \begin{bmatrix} \partial_t & \partial_x & -\partial_y & \partial_z \\ \partial_x & \partial_t & \partial_z & -\partial_y \\ \partial_y & \partial_z & \partial_t & -\partial_x \\ \partial_z & \partial_y & -\partial_x & \partial_t \end{bmatrix} \qquad (13.1)$$

Note that the matrix sum of the inverses of individual variables is not the inverse of the algebraic matrix form.

The $SAS_{L\chi}$ algebra is:

$$SAS_{L\chi} = \begin{bmatrix} a & b & c & d \\ b & a & d & c \\ -c & d & a & -b \\ d & -c & -b & a \end{bmatrix} \qquad (13.2)$$

We choose the matter potential of the A_3 $SAS_{L\chi}$ algebra to be:

$$\Phi_{L\chi}^{SAS} = \begin{bmatrix} \phi & -A_x & -A_y & -A_z \\ -A_x & \phi & -A_z & -A_y \\ A_y & -A_z & \phi & A_x \\ -A_z & A_y & A_x & \phi \end{bmatrix} \tag{13.3}$$

We are guided into choosing this to be the potential (the signs of A_i) by the anticipation that we will extract the anti-symmetric variables to form the electric field and that these anti-symmetric variables must give a space-time curl.

The E-field and B-field are formed as with the quaternions. We have the left-chiral $SAS_{L\chi}$ E-field and the left-chiral $SAS_{L\chi}$ B-field:

$$E_{[1,1]} = \frac{\partial \phi}{\partial t} - \frac{\partial A_x}{\partial x} - \frac{\partial A_y}{\partial y} - \frac{\partial A_z}{\partial z} \qquad B_{[1,1]} = 0$$

$$E_{[1,2]} = \frac{\partial \phi}{\partial x} - \frac{\partial A_x}{\partial t} \qquad B_{[1,2]} = \frac{\partial A_z}{\partial y} + \frac{\partial A_y}{\partial z}$$

$$E_{[1,3]} = -\frac{\partial \phi}{\partial y} - \frac{\partial A_y}{\partial t} \qquad B_{[1,3]} = \frac{\partial A_x}{\partial z} - \frac{\partial A_z}{\partial x} \tag{13.4}$$

$$E_{[1,4]} = \frac{\partial \phi}{\partial z} - \frac{\partial A_z}{\partial t} \qquad B_{[1,4]} = -\frac{\partial A_y}{\partial x} - \frac{\partial A_x}{\partial y}$$

We do the same with the right-chiral A_3 $SAS_{R\chi}$ algebra. The right-chiral $SAS_{R\chi}$ differential operator and $SAS_{R\chi}$ matter potential are:

$$d_{R\chi}^{SAS} = \begin{bmatrix} \partial_t & \partial_x & -\partial_y & \partial_z \\ \partial_x & \partial_t & -\partial_z & \partial_y \\ \partial_y & -\partial_z & \partial_t & \partial_x \\ \partial_z & -\partial_y & \partial_x & \partial_t \end{bmatrix} \qquad \Phi_{R\chi}^{SAS} = \begin{bmatrix} a & -b & -c & -d \\ -b & a & d & c \\ c & d & a & -b \\ -d & -c & -b & a \end{bmatrix} \tag{13.5}$$

This gives the right-chiral $SAS_{R\chi}$ E-field and the right-chiral $SAS_{R\chi}$ B-field as:

$$E_{[1,1]} = \frac{\partial \phi}{\partial t} - \frac{\partial A_x}{\partial x} - \frac{\partial A_y}{\partial y} - \frac{\partial A_z}{\partial z} \qquad B_{[1,1]} = 0$$

$$E_{[1,2]} = \frac{\partial \phi}{\partial x} - \frac{\partial A_x}{\partial t} \qquad B_{[1,2]} = -\frac{\partial A_z}{\partial y} - \frac{\partial A_y}{\partial z}$$

$$E_{[1,3]} = -\frac{\partial \phi}{\partial y} - \frac{\partial A_y}{\partial t} \qquad B_{[1,3]} = -\frac{\partial A_x}{\partial z} + \frac{\partial A_z}{\partial x} \tag{13.6}$$

$$E_{[1,4]} = \frac{\partial \phi}{\partial z} - \frac{\partial A_z}{\partial t} \qquad B_{[1,4]} = \frac{\partial A_y}{\partial x} + \frac{\partial A_x}{\partial y}$$

This right-chiral E-field, (13.6), is identical to the left-chiral E-field above, (13.4). This right-chiral B-field, (13.6), is the reverse of the left-chiral B-field above, (13.4).

Anti-matter fields:

In the quaternion algebras, the real variable can take any value positive or negative. This is because the quaternion algebras form a division algebra without the need to take the exponential of the algebraic matrix form. In the case of the A_3 algebras, these are not division algebras unless we take the exponential. This means the real variable in the algebra, the 'charge' of the algebra, must be positive. We cannot form anti-matter A_3 fields.

The E-fields and the B-fields of the SSA algebras:

The left-chiral *SSA* differential operator and potential are:

$$d_{L\chi}^{SSA} = \begin{bmatrix} \partial_t & \partial_x & \partial_y & -\partial_z \\ \partial_x & \partial_t & \partial_z & -\partial_y \\ \partial_y & -\partial_z & \partial_t & \partial_x \\ \partial_z & -\partial_y & \partial_x & \partial_t \end{bmatrix} \qquad \Phi_{L\chi}^{SSA} = \begin{bmatrix} \phi & -A_x & -A_y & -A_z \\ -A_x & \phi & A_z & A_y \\ -A_y & -A_z & \phi & -A_x \\ A_z & A_y & -A_x & \phi \end{bmatrix} \qquad (13.7)$$

The left-chiral *SSA* E-field and B-field are:

$$E_{[1,1]} = \frac{\partial \phi}{\partial t} - \frac{\partial A_x}{\partial x} - \frac{\partial A_y}{\partial y} - \frac{\partial A_z}{\partial z} \qquad B_{[1,1]} = 0$$

$$E_{[1,2]} = \frac{\partial \phi}{\partial x} - \frac{\partial A_x}{\partial t} \qquad B_{[1,2]} = -\frac{\partial A_z}{\partial y} - \frac{\partial A_y}{\partial z}$$

$$E_{[1,3]} = \frac{\partial \phi}{\partial y} - \frac{\partial A_y}{\partial t} \qquad B_{[1,3]} = \frac{\partial A_x}{\partial z} + \frac{\partial A_z}{\partial x} \qquad (13.8)$$

$$E_{[1,4]} = -\frac{\partial \phi}{\partial z} - \frac{\partial A_z}{\partial t} \qquad B_{[1,4]} = \frac{\partial A_y}{\partial x} - \frac{\partial A_x}{\partial y}$$

The right-chiral *SSA* differential operator and potential are:

$$d_{R\chi}^{SSA} = \begin{bmatrix} \partial_t & \partial_x & \partial_y & -\partial_z \\ \partial_x & \partial_t & -\partial_z & \partial_y \\ \partial_y & \partial_z & \partial_t & -\partial_x \\ \partial_z & \partial_y & -\partial_x & \partial_t \end{bmatrix} \qquad \Phi_{R\chi}^{SSA} = \begin{bmatrix} \phi & -A_x & -A_y & -A_z \\ -A_x & \phi & -A_z & -A_y \\ -A_y & A_z & \phi & A_x \\ A_z & -A_y & A_x & \phi \end{bmatrix} \qquad (13.9)$$

The right-chiral *SSA* E-field and B-field are:

$$E_{[1,1]} = \frac{\partial \phi}{\partial t} - \frac{\partial A_x}{\partial x} - \frac{\partial A_y}{\partial y} - \frac{\partial A_z}{\partial z} \qquad B_{[1,1]} = 0$$

$$E_{[1,2]} = \frac{\partial \phi}{\partial x} - \frac{\partial A_x}{\partial t} \qquad B_{[1,2]} = \frac{\partial A_z}{\partial y} + \frac{\partial A_y}{\partial z}$$

$$E_{[1,3]} = \frac{\partial \phi}{\partial y} - \frac{\partial A_y}{\partial t} \qquad B_{[1,3]} = -\frac{\partial A_x}{\partial z} - \frac{\partial A_z}{\partial x}$$

$$E_{[1,4]} = -\frac{\partial \phi}{\partial z} - \frac{\partial A_z}{\partial t} \qquad B_{[1,4]} = -\frac{\partial A_y}{\partial x} + \frac{\partial A_x}{\partial y}$$

(13.10)

The E-fields and the B-fields of the ASS algebras:
The left-chiral ASS differential operator and potential are:

$$d_{L\chi}^{ASS} = \begin{bmatrix} \partial_t & -\partial_x & \partial_y & -\partial_z \\ \partial_x & \partial_t & -\partial_z & \partial_y \\ \partial_y & -\partial_z & \partial_t & \partial_x \\ \partial_z & \partial_y & -\partial_x & \partial_t \end{bmatrix} \qquad \Phi_{L\chi}^{ASS} = \begin{bmatrix} \phi & -A_x & -A_y & -A_z \\ A_x & \phi & A_z & -A_y \\ -A_y & A_z & \phi & A_x \\ -A_z & -A_y & -A_x & \phi \end{bmatrix}$$

(13.11)

The left-chiral ASS E-field and B-field are:

$$E_{[1,1]} = \frac{\partial \phi}{\partial t} - \frac{\partial A_x}{\partial x} - \frac{\partial A_y}{\partial y} - \frac{\partial A_z}{\partial z} \qquad B_{[1,1]} = 0$$

$$E_{[1,2]} = -\frac{\partial \phi}{\partial x} - \frac{\partial A_x}{\partial t} \qquad B_{[1,2]} = \frac{\partial A_z}{\partial y} - \frac{\partial A_y}{\partial z}$$

$$E_{[1,3]} = \frac{\partial \phi}{\partial y} - \frac{\partial A_y}{\partial t} \qquad B_{[1,3]} = -\frac{\partial A_x}{\partial z} - \frac{\partial A_z}{\partial x}$$

$$E_{[1,4]} = \frac{\partial \phi}{\partial z} - \frac{\partial A_z}{\partial t} \qquad B_{[1,4]} = \frac{\partial A_y}{\partial x} + \frac{\partial A_x}{\partial y}$$

(13.12)

The right-chiral ASS differential operator and potential is:

$$d_{R\chi}^{ASS} = \begin{bmatrix} \partial_t & -\partial_x & \partial_y & \partial_z \\ \partial_x & \partial_t & \partial_z & -\partial_y \\ \partial_y & \partial_z & \partial_t & -\partial_x \\ \partial_z & -\partial_y & \partial_x & \partial_t \end{bmatrix} \qquad \Phi_{R\chi}^{ASS} = \begin{bmatrix} \phi & -A_x & -A_y & -A_z \\ A_x & \phi & -A_z & A_y \\ -A_y & -A_z & \phi & -A_x \\ -A_z & A_y & A_x & \phi \end{bmatrix}$$

(13.13)

The right-chiral ASS E-field and B-field is:

$$E_{[1,1]} = \frac{\partial \phi}{\partial t} - \frac{\partial A_x}{\partial x} - \frac{\partial A_y}{\partial y} - \frac{\partial A_z}{\partial z} \qquad\qquad B_{[1,1]} = 0$$

$$E_{[1,2]} = -\frac{\partial \phi}{\partial x} - \frac{\partial A_x}{\partial t} \qquad\qquad B_{[1,2]} = -\frac{\partial A_z}{\partial y} + \frac{\partial A_y}{\partial z}$$

$$E_{[1,3]} = \frac{\partial \phi}{\partial y} - \frac{\partial A_y}{\partial t} \qquad\qquad B_{[1,3]} = \frac{\partial A_x}{\partial z} + \frac{\partial A_z}{\partial x}$$

$$E_{[1,4]} = \frac{\partial \phi}{\partial z} - \frac{\partial A_z}{\partial t} \qquad\qquad B_{[1,4]} = -\frac{\partial A_y}{\partial x} - \frac{\partial A_x}{\partial y}$$

$$(13.14)$$

The emergent spaces:

The *SAS* emergent E-field is non-chiral; that E-field is:

$$E_{SAS}^{Emergent} = \begin{bmatrix} Div & \dfrac{\partial \phi}{\partial x} - \dfrac{\partial A_x}{\partial t} & -\dfrac{\partial \phi}{\partial y} - \dfrac{\partial A_y}{\partial t} & \dfrac{\partial \phi}{\partial z} - \dfrac{\partial A_z}{\partial t} \\[2ex] \dfrac{\partial \phi}{\partial x} - \dfrac{\partial A_x}{\partial t} & Div & 0 & 0 \\[2ex] \dfrac{\partial \phi}{\partial y} + \dfrac{\partial A_y}{\partial t} & 0 & Div & 0 \\[2ex] \dfrac{\partial \phi}{\partial z} - \dfrac{\partial A_z}{\partial t} & 0 & 0 & Div \end{bmatrix} \qquad (13.15)$$

$$Div = \frac{\partial \phi}{\partial t} - \frac{\partial A_x}{\partial x} - \frac{\partial A_y}{\partial y} - \frac{\partial A_z}{\partial z}$$

The *SAS* emergent B-field is chiral; that B-field is:

$$B_{SAS}^{Emergent} = \begin{bmatrix} 0 & 0 & 0 & 0 \\[2ex] 0 & 0 & -\dfrac{\partial A_y}{\partial x} - \dfrac{\partial A_x}{\partial y} & \dfrac{\partial A_x}{\partial z} - \dfrac{\partial A_z}{\partial x} \\[2ex] 0 & -\dfrac{\partial A_y}{\partial x} - \dfrac{\partial A_x}{\partial y} & 0 & -\dfrac{\partial A_z}{\partial y} - \dfrac{\partial A_y}{\partial z} \\[2ex] 0 & -\dfrac{\partial A_x}{\partial z} + \dfrac{\partial A_z}{\partial x} & -\dfrac{\partial A_z}{\partial y} - \dfrac{\partial A_y}{\partial z} & 0 \end{bmatrix} \qquad (13.16)$$

There are similar results for the other two pairs of A_3 algebras.

There are six A_3 algebras. We form the emergent E-field and the emergent B-field by adding the six E-fields and the six B-fields of each algebra, (13.4) & (13.6) & (13.8) & (13.10) & (13.12) & (13.14).

We form the symmetric part and the anti-symmetric part of the emergent field. Each A_3 algebra has three imaginary variables. Of the three imaginary variables, two are symmetric across the leading diagonal and one is anti-symmetric across the leading diagonal.

The anti-symmetric emergent A_3 E-field is:

$$2\begin{bmatrix} 3Div & -\dfrac{\partial \phi}{\partial x}-\dfrac{\partial A_x}{\partial t} & -\dfrac{\partial \phi}{\partial y}-\dfrac{\partial A_y}{\partial t} & -\dfrac{\partial \phi}{\partial z}-\dfrac{\partial A_z}{\partial t} \\[2mm] \dfrac{\partial \phi}{\partial x}+\dfrac{\partial A_x}{\partial t} & 3Div & 0 & 0 \\[2mm] \dfrac{\partial \phi}{\partial y}+\dfrac{\partial A_y}{\partial t} & 0 & 3Div & 0 \\[2mm] \dfrac{\partial \phi}{\partial z}+\dfrac{\partial A_z}{\partial t} & 0 & 0 & 3Div \end{bmatrix} \qquad (13.17)$$

$$Div = \frac{\partial \phi}{\partial t} - \frac{\partial A_x}{\partial x} - \frac{\partial A_y}{\partial y} - \frac{\partial A_z}{\partial z}$$

The anti-symmetric emergent A_3 B-field is:

$$2\begin{bmatrix} 0 & 0 & 0 & 0 \\[2mm] 0 & 0 & -\dfrac{\partial A_y}{\partial x}+\dfrac{\partial A_x}{\partial y} & \dfrac{\partial A_x}{\partial z}-\dfrac{\partial A_z}{\partial x} \\[2mm] 0 & \dfrac{\partial A_y}{\partial x}-\dfrac{\partial A_x}{\partial y} & 0 & -\dfrac{\partial A_z}{\partial y}+\dfrac{\partial A_y}{\partial z} \\[2mm] 0 & -\dfrac{\partial A_x}{\partial z}+\dfrac{\partial A_z}{\partial x} & \dfrac{\partial A_z}{\partial y}-\dfrac{\partial A_y}{\partial z} & 0 \end{bmatrix} \qquad (13.18)$$

Taken together as a sum, the A_3 emergent anti-symmetric E-field with zero divergence and the A_3 emergent anti-symmetric B-field are the electromagnetic tensor, (13.17) & (13.18).

Similar considerations produce the emergent symmetric tensor which is the energy momentum tensor of general relativity.

A second non-commutative differentiation of the above anti-symmetric E-fields and anti-symmetric B-fields produces the Maxwell equations of classical electromagnetism[80],[81]

Parity (handedness):
Within the A_3 algebras, (13.4) & (13.6), we have:

$$E_{L\chi}^{Matter} = E_{R\chi}^{Matter} \qquad (13.19)$$

We see that the A_3 E-field is 'immune' to chirality.

We also have within the A_3 algebras:

---

[80] See: Dennis Morris : Upon General Relativity.
[81] See: Dennis Morris : The Physics of Empty Space.

$$B_{L\chi}^{Matter} = -B_{R\chi}^{Matter} \tag{13.20}$$

We see that the A_3 B-field changes with chirality. This is spatial parity. The B-field is reversed under a spatial parity transformation. We have discovered spatial parity.

In these A_3 spaces, chirality is just handedness, which we also call spatial parity. The absence of anti-matter from these A_3 spaces means that the chirality is less complicated than in the quaternion spaces.

Conclusion:

We have come to the understanding that each space has its own form of chirality. We have come to the view that, as a consequence of chirality being expressed in the distribution of the minus signs within a non-commutative algebraic matrix form, chirality is also expressed in the E-fields and the B-fields of an algebra and in the commutation relations of the algebra.

Because the real variable in the quaternion algebras can take both negative values and positive values, the quaternion spaces have anti-matter E-fields and anti-matter B-fields. Because the real variable in the A_3 algebras can take only positive values, the A_3 spaces have no anti-matter E-fields and no anti-matter B-fields.

Anti-matter in our Universe

Our 4-dimensional space-time is the emergent space of the six A_3 algebras. There is no anti-matter in the emergent space of the six A_3 algebras because these algebras are division algebras for only positive values of the real variable. Because there is no anti-matter in the emergent space of the six A_3 algebras, there is no anti-matter in our 4-dimensional space-time.

There is anti-matter in the quaternion emergent space. We must therefore presume that any anti-matter within our universe is within the quaternion space. Thus, we must presume that our universe is comprised of at least the A_3 emergent space, our 4-dimensional space-time, and the quaternion emergent space. We have at least two types of space in our universe. This fits with the observation of at least two kinds of chirality in our universe.

We see immediately that there is likely to be less anti-matter in our universe than matter. Indeed, it seems that the only source of anti-matter will be a 'leak' from the quaternion space into our 4-dimensional space-time. It seems as though the particle reactions which create anti-matter happen within quaternion space where conservation of energy and momentum are not physical laws – off mass-shell - but where conservation of electric charge and intrinsic angular momentum are physical laws. All of this fits with observation.

The nature of the particle/anti-particle relationship is the chirality of quaternion space. This is different from the chirality of our 4-dimensional space-time which is spatial parity. The quaternion space chirality comes in two forms which are the E-field chirality, which is what we normally associate with anti-matter and call charge conjugation, and the B-field chirality which is CP invariance. Perhaps we need to rethink what we mean by anti-matter.

Chirality and Rotation and Intrinsic Spin

Consider the product of two left-chiral quaternion rotation matrices in which two of the variables are zero and the other variable is $\dfrac{\pi}{2}$. We have:

$$\begin{bmatrix} \cos\dfrac{\pi}{2} & \sin\dfrac{\pi}{2} & 0 & 0 \\ -\sin\dfrac{\pi}{2} & \cos\dfrac{\pi}{2} & 0 & 0 \\ 0 & 0 & \cos\dfrac{\pi}{2} & -\sin\dfrac{\pi}{2} \\ 0 & 0 & \sin\dfrac{\pi}{2} & \cos\dfrac{\pi}{2} \end{bmatrix} \begin{bmatrix} \cos\dfrac{\pi}{2} & 0 & \sin\dfrac{\pi}{2} & 0 \\ 0 & \cos\dfrac{\pi}{2} & 0 & \sin\dfrac{\pi}{2} \\ -\sin\dfrac{\pi}{2} & 0 & \cos\dfrac{\pi}{2} & 0 \\ 0 & -\sin\dfrac{\pi}{2} & 0 & \cos\dfrac{\pi}{2} \end{bmatrix} = \begin{bmatrix} 0 & 0 & 0 & 1 \\ 0 & 0 & -1 & 0 \\ 0 & 1 & 0 & 0 \\ -1 & 0 & 0 & 0 \end{bmatrix} \quad (15.1)$$

The two left-chiral rotation matrices can be thought of as denoting points, position vectors, in left-chiral quaternion space. The left-most matrix is a unit position vector pointing in the b direction; the other matrix is a unit position vector pointing in the c direction. The result of combining the two quaternion rotations is the product matrix which is a unit position vector pointing in the plus d direction.

Let us do the same with two right-chiral quaternion rotation matrices:

$$\begin{bmatrix} \cos\dfrac{\pi}{2} & \sin\dfrac{\pi}{2} & 0 & 0 \\ -\sin\dfrac{\pi}{2} & \cos\dfrac{\pi}{2} & 0 & 0 \\ 0 & 0 & \cos\dfrac{\pi}{2} & \sin\dfrac{\pi}{2} \\ 0 & 0 & -\sin\dfrac{\pi}{2} & \cos\dfrac{\pi}{2} \end{bmatrix} \begin{bmatrix} \cos\dfrac{\pi}{2} & 0 & \sin\dfrac{\pi}{2} & 0 \\ 0 & \cos\dfrac{\pi}{2} & 0 & -\sin\dfrac{\pi}{2} \\ -\sin\dfrac{\pi}{2} & 0 & \cos\dfrac{\pi}{2} & 0 \\ 0 & \sin\dfrac{\pi}{2} & 0 & \cos\dfrac{\pi}{2} \end{bmatrix} = \begin{bmatrix} 0 & 0 & 0 & -1 \\ 0 & 0 & -1 & 0 \\ 0 & 1 & 0 & 0 \\ 1 & 0 & 0 & 0 \end{bmatrix} \quad (15.2)$$

The two right-chiral rotation matrices can be thought of as denoting points, position vectors, in right-chiral quaternion space. The left-most matrix is a unit position vector pointing in the b direction; the other matrix is a unit position vector pointing in the c direction. The result of combining the two quaternion rotations is the product matrix which is a unit position vector pointing in the minus d direction.

Intrinsic spin:

Conventionally, intrinsic spin, is thought of as having an axis of rotation just as classical rotation has an axis of rotation. However, utterly unlike classical rotation, the axis of intrinsic spin cannot point in any direction.

It is extensively verified that the 'axis' of intrinsic spin of an electron points in one of only two directions with respect to whatever other vector the intrinsic spin of the electron is being measured. This is why we have precise spectral lines and precise fine splitting of these precise spectral lines in atoms rather than a thick smudge of a spectral line.

The electron, and the neutrino and other sub-atomic particles, are point particles of zero spatial extent, and so these sub-atomic particles cannot rotate in the classical sense, and yet they have intrinsic spin.

Intrinsic spin is chirality of commutation relations:

In this book, we will opine that intrinsic spin is chirality within a spinor space like quaternion space. In quaternion space, there is no axis of rotation, but there are two chiralities, 'directions of rotation', determined by the commutation relations of the two types of quaternions. Let us imagine that an electron is really a quaternion; then there are two types of electron. One type of electron is left-chiral, and the other type of electron is right-chiral. We thus associate the two types of intrinsic spin, up or down, of sub-atomic particles like electrons and neutrinos with the two types of quaternion commutation relations.

There is a fly in the ointment. Intrinsic angular momentum does manifest itself as classical angular momentum. This is shown quite clearly by the Einstein de Haas effect[82]. By some mechanism, intrinsic angular momentum with no spatial extent becomes classical angular momentum associated with spatial extent. We do not properly understand how this happens.

Superposition of electron fields:

In this book, we see the electron as a superposition of the E-fields of the two quaternion types:

$$e^- = x\left|\mathbb{H}_{L\chi}\right\rangle_{E-field} + y\left|\mathbb{H}_{R\chi}\right\rangle_{E-field} \tag{15.3}$$

Wherein $\{x, y\}$ are the relative amplitudes (amounts) of the different types of quaternion.

Since electrons move at a velocity which is less than the speed of light, we can find two observers whose velocities are such that one observer sees the electron moving from right to left and the other observer sees the electron moving from left to right. When looking at the same electron, one observer will say the electron has its intrinsic spin aligned with the electron's velocity and the other observer will say the electron has its intrinsic spin aligned oppositely to the electron's velocity.

---

[82] The Einstein de Haas effect was predicted by O. W. Richardson in 1908 – Phys. Review 26 248 (1908) and verified by J. Q. Stewart in 1918 – Phys. Review 11 100 (1918) and by A. P. Chattock and L. F. Bates in 1923 – Phil Trans. Roy. Soc. A 223 257 (1923).

Summary:

We opine that that intrinsic spin comes in two directions corresponding to the two chiralities of the quaternions.

The reader is warned that the way we opine is not the contemporary conventional view. The contemporary conventional view is that intrinsic spin is rotation about an axis and the fact that the electron is a point of zero extent is 'swept under the carpet'.

Chapter 16

CP Violation and the Weird Nature of the Kaon

In October 1946 and again in May 1947, the British physicists George Dixon Rochester (1908-2001) and Clifford Charles Butler (1922-1999) observed in cloud chamber photographs a particle which was roughly a thousand times the mass of an electron. These particles formed V shaped tracks in the cloud chamber, and so they were initially called V particles. Over the next six years, more V particles were observed in cosmic ray experiments. From observation of the particles produced by the decay of the V particles, it became apparent that there were two types of V particles. One type of V particle always decayed into a set of particles which included a proton; these V particles were subsequently named hyperons. The other type of V particle always decayed into only mesons; these V particles were subsequently named K-mesons and the name mutated into kaons. Kaon or K-meson – both the same thing. The physicists of the time found the behaviour of the hyperons and the kaons to be strange, and thus these V particles, hyperons and kaons, became referred to as 'strange particles'.

Strange particles:
The strange aspect of the behaviour of the hyperons and the kaons is that they take much longer to decay than would be expected. The speed of the decay of a sub-atomic particle is associated with the type of nuclear force which is involved in the decay. If the strong nuclear force is involved in a particle decay, then that particle will typically have a half-life of circa 10^{-23} seconds. If the weak nuclear force is involved in a particle decay, then that particle will typically have a half-life of circa 10^{-10} seconds; this is ten trillion times longer than the strong force decay half-life. The 'weakness' of the weak nuclear force as demonstrated in decay times compared to the 'strength' of the strong nuclear force as demonstrated in decay times is the reason these two forces are called the weak force and the strong force.

It seemed to physicists at the time that the strange particles were produced in interactions which involved the strong nuclear force. They were correct; since this is the case, we would expect the strange particles to decay 'via the strong force'[83] and thus have half-lives of about 10^{-23} seconds. The strange particles have half-lives of circa 10^{-10} seconds; this indicates that they decay 'via the weak force'.

In 1952, the American physicist Abraham Pais (1918-2000) suggested that the strange particles, hyperons and kaons, could not be produced one-at-a-time by the strong interaction[84] but can be produced only in pairs. This was found to be correct in 1953 at the Brookhaven accelerator. Strange particles are produced in reactions like:

$$\pi^- + p \rightarrow \Lambda^0 + K^0 \tag{16.1}$$

---

[83] The phrase 'via the strong force' or 'via the weak force' is common parlance in particle physics. The phases indicate which of the forces, weak or strong, is involved in the particluar particle interaction being described.

[84] It is another bit of the esoteric phraseology of particle physics that the strong force is often called the strong interaction. Similarly, the weak force is often called the weak interaction.

wherein the π^- is a negatively charged pion (pi-meson), the p is a proton, the Λ^0 is an electrically neutral hyperon, and the K^0 is an electrically neutral kaon[85]; the superscript $\{+,-,0\}$ indicates the electrical charge of the particle. The importance of strange particles being produced only in pairs is that it implies there is some kind of 'charge' possessed by the strange particles which is conserved by the strong force in particle reactions. It was this conservation of a type of 'charge' that was proposed by Pais. We now call this 'charge' strangeness. The proposal was later supported by Murray Gell-Mann (1929-) and Kazuhiko Nishijima (1926-2009). We assign the hyperon an amount of strangeness charge of minus one, and we assign the kaon an amount of strangeness charge of plus one. Because the pi-meson and the proton are not strange particles, they have zero strangeness charge. Looking at the reaction, (16.1), we see that the total amount of strangeness charge before the reaction is zero and the total amount of strangeness charge after the reaction is also zero:

$$\pi^- + p \rightarrow \Lambda^0 + K^0$$
$$0 \; + \; 0 \rightarrow (-1) + (+1) \tag{16.2}$$

Strangeness charge is conserved. Actually, strangeness charge is conserved by the strong nuclear force, strong interaction. Thus, a single strange particle, like a hyperon or a kaon, cannot decay via the strong force because this would require violating conservation of strangeness charge. We tend to say just strangeness rather than strangeness charge.

Of course, as said above, single strange particles do decay. A typical decay is of the form:

$$\Lambda^0 \rightarrow_{\Delta s=-1} \pi^- + p \tag{16.3}$$

Is this decay, (16.3), minus one unit of strangeness is lost. This decay, (16.3), violates conservation of strangeness. This decay, (16.3), cannot be via the strong force because the strong force respects conservation of strangeness. The decay, (16.3), has a half-life of 10^{-10} seconds; it is a decay via the weak force. The weak force does not respect conservation of strangeness. If all interactions respected conservation of strangeness, then a single kaon or a single hyperon could never decay into particles which have zero strangeness.

An important point is that the different types of force, different types of interaction, obey different conservation laws. The strong force is associated with conservation of strangeness. The weak force does not 'feel' strangeness, and so the weak force does not conserve strangeness. None-the-less, there is a complication. The weak interaction can 'not feel' strangeness in only one unit at a time.

We might think that, since the weak interaction, weak force, does not 'feel' strangeness, then a particle with two units of strangeness would be able to decay via the weak force in a single reaction. There is a type of hyperon called the xi hyperon[86] which has strangeness charge of minus two. The xi hyperon decays via two weak force reactions never via one weak force reaction. We have:

$$\Xi^0 \rightarrow_{\Delta S=-1} \Lambda^0 + \pi^0 \rightarrow_{\Delta S=-1} \pi^- + p + \pi^0 \tag{16.4}$$

---

[85] As with all particle interactions, electric charge is always conserved.
[86] The xi hyperon is also called the cascade particle.

Kaons:

Kaons are mesons. Today, we understand that mesons are comprised of a quark and an anti-quark. There are six types of quarks[87]:

$$
\begin{array}{cccccc}
up & u & charm & c & top & t \\
down & d & strange & s & bottom & b
\end{array}
\qquad (16.5)
$$

The quarks on the top row have electric charge of $+\dfrac{2}{3}$, and the quarks on the bottom row have electric charge of $-\dfrac{1}{3}$. To each of these quarks, there is a corresponding anti-quark whose electric charge is the reverse of the electric charge of the quark.

The kaons are comprised of up quarks, down quarks, and strange quarks. Thus, there are four possible ways of combining up quarks, down quarks, and strange quarks to make kaons; these ways are:

$$
K^+ = u\bar{s} \qquad K^- = \bar{u}s \qquad K^0 = d\bar{s} \qquad \overline{K^0} = \bar{d}s \qquad (16.6)
$$

These are the four types of kaon. We see that these mesons have electrical charge of $\{+1, -1, \text{ or } 0\}$. All mesons, and indeed all possible particles which are a combination of quarks, always have electrical charge of $\{+1, -1, \text{ or } 0\}$. Clearly, there are other strange mesons to be formed from a strange quark and one of the $\{c, t, b\}$ quarks, but this does not presently concern us.

Pions:

By the way, the pairs of $\{u, d\}$ quarks are pions (pi-mesons):

$$
\pi^+ = u\bar{d} \qquad \pi^- = \bar{u}d \qquad \pi^0 = u\bar{u} \qquad \pi^0 = \bar{d}d \qquad (16.7)
$$

Pions have zero intrinsic spin. The masses of the pions are:

$$
\pi^+_{mass} = \pi^-_{mass} = 139.57018 \pm 0.00035 \ Mev/c^2 \qquad\qquad \pi^0_{mass} = \pi^0_{mass} = 134.9766 \pm 0.0006 \ Mev/c^2
$$

$$(16.8)$$

The π^+ particle is taken to be the anti-particle of the π^- particle; looking at the quark structures of these two particles, that makes sense. The two π^0 particles are taken to be the anti-particles of each other; looking at the quark structures of the two π^0 particles, the assertion that these are anti-particles of each other seems not to hold water. The assertion is made to fit with a scheme based on the particles having something called isospin; perhaps we need to question this isospin stuff.

---

[87] Each type of quark comes in three varieties called colours, red, green, and blue.

Back to kaons:

The K^+ & $K^0 = d\bar{s}$ kaons are assigned strangeness of +1. The K^- & $\overline{K^0} = \bar{d}s$ kaons are assigned strangeness of −1. The masses of the kaons are:

$$K^+_{mass} = K^-_{mass} = 493.667 \pm 0.013 \ Mev/c^2 \qquad\qquad K^0_{mass} = \overline{K^0}_{mass} = 497.648 \pm 0.022 \ Mev/c^2$$

$$(16.9)$$

The intrinsic spin of the kaons is zero; this fits with the intrinsic spins of the constituent quarks $\left\{ +\frac{1}{2}, -\frac{1}{2} \right\}$ cancelling each other. Looking at the masses and quark structures of the kaons, we see that K^+ is the anti-particle of K^-, and this is conventionally taken to be the correct understanding. However, in spite of the coincident masses and the quark structures, the two K^0 particles are conventionally taken not to be anti-particles to each other; this also is to do with the isospin stuff.

Intrinsic parity:

In our 4-dimensional space-time, we can flip the spatial parity of something, say a wave function, by changing the sign of the three spatial co-ordinates. This is the spatial reflection, reflection in a mirror, with which we are all familiar. Within particle physics, there is a property of particles which is called intrinsic parity. Individual particles are assigned an intrinsic parity of plus one or minus one. Thus, intrinsic parity is a quantum number attached to a particle as part of the identity of that particle. Remarkably, this intrinsic parity quantum number seems to manifest itself within our 4-dimensional space-time as geometric spatial parity – hence the name. Particles with intrinsic parity of plus one 'just are', but particles with intrinsic parity of minus one are mirror reflections of the 'just are' particles.

Within particle physics, although the intrinsic parity of a particle is meaningful to the strong interaction, the electromagnetic interaction, and the gravitational interaction, gravity, intrinsic parity is meaningless to the weak interaction. The weak force does not respect intrinsic parity. The weak force does not 'feel' the parity of a particle.

It is a little like a colour-blind person does not see any difference between a blue snooker ball and a red snooker ball. The way that colour-blind people keep the score at snooker is different from the way that normally sighted people keep the score at snooker – indeed, the whole game is different. So it is that 'the whole game is different' when particles are subject to the weak force from when those same particles are subject to the strong force. The remarkable bit of all this is that the weak force is 'blind' to the spatial reflection parity of our 4-dimensional space-time.

Kaon decay and parity:

In 1954, William Chinowsky and Jack Steinberger (1921-) were able to demonstrate that all pions have intrinsic parity of minus one[88]. In the early 1950's, physicists observed two distinct types of decay products seemingly associated with two previously unknown particles:

---

[88] Chinowsky W, Steinberger J. (1954) Absorption of Negative Pions in Deuterium. Parity of the Pion : Physical Review 95 (6) 1561-1564.

$$?^+ \rightarrow \pi^+ + \pi^+ + \pi^- \qquad \& \qquad ?^+ \rightarrow \pi^+ + \pi^0 \qquad (16.10)$$

The left-hand decay product of (16.10) has an intrinsic parity of minus three. The right-hand decay product of (16.10) has an intrinsic parity of minus two. Because these two sets of decay products were of different parity, it was assumed that these two sets of decay products must be from the decay of two different particles originally named the τ^+ and the θ^+ particles. It was later realised that in fact both the sets of decay products, (16.10), derive from the decay of a K^+ kaon. Both decays correspond to half-lives of circa 10^{-8} seconds which is typical of the weak force. Clearly, the weak force must not respect conservation of intrinsic parity. Because of the connection between spatial parity in our 4-dimensional space-time and intrinsic parity, the violation of intrinsic parity by the weak force implies a violation of spatial parity in our 4-dimensional space-time by the weak force – physics in a left-handed co-ordinate system is different from physics in a right-handed co-ordinate system.

"Well, that c'ant possibly be true[89]", they all said. In 1956, the two Chinese physicists Tsung-Dao Lee (1926-) and Chen Ning Yang (1922-) realised that there was no experimental evidence that the weak force respected the conservation of intrinsic parity[90]. Lee and Yang predicted the violation of intrinsic parity, and thus the violation of spatial parity, by the weak force.[91]

Experimental verification of violation of parity:

In 1956, it was shown experimentally at the National Bureau of Standards in Washington by the Chinese physicist Madam Chien-Siung Wu[92] that the weak force violates intrinsic parity. The weak force distinguishes between right-handed physics and left-handed physics. The experiment conducted by Wu observed the spatially asymmetric emission of β-decay electrons from radioactive cobalt in the weak force reaction:

$$Co^{60} \rightarrow Ni^{60} + e^- + \bar{\nu} \qquad (16.11)$$

The violation of intrinsic parity by the weak force was seen to correspond to a spatial violation of parity because the electrons emitted by the decay were not emitted in equal numbers in all possible directions within our 4-dimensional space-time.

CP invariance:

Physicists like conservation laws. There is something absolutely reliable about a conservation law. By the 1960's, physicists had accepted that, just as strangeness is not conserved by the weak force, neither is intrinsic parity conserved by the weak force. Further, observations of the intrinsic spin of electrons and positrons emitted in the decays of electrically positively charged and electrically negatively charged muons showed that the weak force does not respect charge-conjugation symmetry, but it did seem that a combined symmetry of charge-conjugation and parity would be conserved by the weak force. The

---

[89] Physicists were so confused by all this that, unthinkingly, they even got their apostophies in the wrong places by reflecting them from one end of the word to the other.

[90] Lee T. D. Yang C. N. (Oct 1956) Question of Parity Conservation in Weak Interactions : Physical Review 104 (1) 254

[91] This prediction by Tsung-Dao Lee and Chen Ning Yang won them the Nobel prize in 1957.

[92] Wu C. S. Ambler E Hayward R. W. Hoppes D. D. Hudson R. P. (1957) Experimental test of Parity Conservation in beta decay. Physical Review 105(4) 1413-1415

conservation of charge-conjugation and parity is called CP invariance. CP invariance is that right-handed anti-matter physics is the same as left-handed matter physics. "Well, at least the weak force respects something", the physicists all said. They spoke too early.

Violation of CP invariance:

In 1964, James Cronin (1931-2016) and Val Logsdon Fitch (1923-2015), in what is now known as the Fitch Cronin experiment, discovered the violation of CP invariance, CP violation, in the decay of neutral kaons[93]. Cronin and Fitch were awarded the 1980 Nobel Prize in Physics for their discovery. Searches for other examples of the violation of CP invariance throughout the 1960's, 1970's and 1980's found no other examples of violation of CP invariance until 1989 when the NA31 experiment at CERN also suggested CP invariance was violated for kaons. The result of the NA31 experiment was confirmed in 1999 by the KTev experiment[94] at Fermilab and the NA48 experiment[95] at CERN.

In 2001, the BaBar experiment at the Stanford Linear Accelerator Centre, SLAC, observed violation of CP invariance in B-mesons[96]. These violations were confirmed by the Belle experiment at the High Energy Accelerator Research Organisation, KEK, in Japan[97]. In 2013, the LHCb discovered CP violation in strange B-meson decays; this was confirmed by the BaBar and Belle experiments in 2015.

It seems that the weak force does not 'see' the neutral kaons, K^0, in the same way that the strong force 'sees' the neutral kaons. Technically, the two neutral kaons are eigenstates of the parity operator but they are not eigenstates of the CP operator.

An important point for our purposes is that, although violation of CP invariance has been observed in the quark sector –mesons, it has not been observed in the electron and neutrino sector.

Anti-matter and the big bang:

According to our present understanding of particle physics and the big bang origin of the universe, there should be as much anti-matter in the universe as there is matter unless CP invariance is violated. The discovery of CP violation by Cronin and Fitch was initially welcomed by cosmologists because it allows the preponderance of matter over anti-matter within our universe to be explained without having to arbitrarily assume such a preponderance at the start of the universe. Unfortunately, it is now understood that there is not enough CP violation to account for the preponderance of matter over anti-matter. Particle physicists are therefore looking for more instances of CP violation.

[93] J. H. Christenson, J. W. Cronin, V. L. Fitch, & R. Turley (1964) Evidence for the 2π decay of the K_2^0 meson system : Physical Review Letters 13 138

[94] Observation of direct CP violationin $K_{S,L} \rightarrow \pi\pi$ decays : Physical Review Letters 83 22 1999.

[95] NA48 Collaboration V. Fanti, A. Lai, D. Marras, L. Musa et al (1999) A new measurement of direct CP violation in the two pion decays of the neutral kaon : Physics Letters B 465 (1-4) 335-348.

[96] Measurements of CP violating assymetries in B^0 decays to CP eigenstates: Physical Review Letters 86 2515 2001.

[97] Observations of large CP violation in the Neutral B-meson system: Physical Review Letters 87 2001.

The CPT theorem:

The CPT theorem states that the combined symmetries of charge-conjugation, swapping matter for anti-matter, spatial parity, swapping right-hand for left-hand, and time reversal is conserved and respected by all physics. Cronin and Fitch discovered that right-handed anti-matter physics is different from left-handed matter physics, but that applies only if both right-handed anti-matter physics and left-handed matter physics are both done with time moving in the same direction. The CPT theorem says that right-handed anti-matter physics moving forwards in time is the same as left-handed matter physics moving backwards in time.

There is no experimental evidence of CPT conservation being violated.

If the CPT theorem is correct, then violation of CP invariance implies that physics moving forwards in time is not the same as physics moving backwards in time. Given that we cannot move backwards in time, what does this mean?

The expanding universe will be larger an hour from now than it was an hour before now, but this is not what is meant by forward time physics being different from backward time physics.

When we get to the 8-dimensional $C_2 \times C_2 \times C_2$ algebras, we will discover that each of these algebras has a variable, the e variable, which 'mimics' the time variable in many ways. In the 8-dimensional algebras, the time variable must be positive because we need to take the exponential of the algebraic matrix form to form a division algebra. In the 8-dimensional algebras, the e variable can be positive or negative. Perhaps this is connected to time reversal physics. In the 8-dimensional algebras, we will see that CP invariance is violated.

Preamble to the 8-dimensional Algebras

Chirality does not exist within the 2-dimensional division algebras, $\mathbb{C}\,\&\,\mathbb{S}$ because these are commutative algebras. We have an understanding of chirality within the six non-commutative 4-dimensional A_3 algebras and within the two non-commutative 4-dimensional the quaternion algebras which derive from the $C_2 \times C_2$ finite group. There are no other 4-dimensional chiral algebras, and so we have covered chirality in all its 4-dimensional forms.

We understand that only non-commutative algebras from commutative finite groups hold chirality. Such finite groups are rare. The order eight $C_2 \times C_2 \times C_2$ finite group is a commutative finite group which holds non-commutative algebras; it is also the only finite group other than the $C_2 \times C_2$ group that holds chiral algebras which can be manifest in our 4-dimensional space-time.

The next step in our exploration of algebraic chirality is to examine the algebras of the $C_2 \times C_2 \times C_2$ group.

The algebras of the $C_2 \times C_2 \times C_2$ group are the 8-dimensional Clifford algebras. As division algebras, the 8-dimensional Clifford algebras are the division algebras that are derived from the commutative finite group $C_2 \times C_2 \times C_2$[98]. Our interest in these algebras is based upon the anticipation that these algebras might be connected to the quarks and the strong nuclear force.

In this book, we use the term '8-dimensional Clifford algebra' to mean the same as the non-commutative algebras which derive from the $C_2 \times C_2 \times C_2$ finite group.

Lots of algebras:
In total, there are one thousand and twenty-four 8-dimensional Clifford algebras that derive from the finite group $C_2 \times C_2 \times C_2$. These 1024 algebras are divided into eight sets of one hundred and twenty-eight algebras.

An 8-dimensional Clifford algebra:
The traditional 8-dimensional Clifford algebras are not division algebras because they contain zero divisors[99]. These traditional Clifford algebras are denoted by $Cl_{i,j}$ where the subscripts, $i \,\&\, j$, are the number of plus signs and the number of minus signs respectively in the quadratic form distance function

---

[98] See: Dennis Morris : The Naked Spinor
[99] Any algebra which has a square root of plus unity has zero divisors. We rid ourselves of the zero divisors by taking the exponential of the algebra and thereby form a division algebra.

used to derive the Clifford algebra; for example, the $Cl_{2,1}$ algebra is derived from the $(+,+,-)$ distance function:

$$dist^2 = x^2 + y^2 - z^2 \tag{17.1}$$

Using:

$$\left(x\vec{e_1} + y\vec{e_2} - z\vec{e_3}\right)\left(x\vec{e_1} + y\vec{e_2} - z\vec{e_3}\right) = x^2 + y^2 - z^2 \tag{17.2}$$

We get:

$$x^2\vec{e_1}\vec{e_1} + xy\vec{e_1}\vec{e_2} - xz\vec{e_1}\vec{e_3} + xy\vec{e_2}\vec{e_1} + y^2\vec{e_2}\vec{e_2} - yz\vec{e_2}\vec{e_3} - xz\vec{e_3}\vec{e_1} - yz\vec{e_3}\vec{e_2} + z^2\vec{e_3}\vec{e_3} = x^2 + y^2 - z^2 \tag{17.3}$$

The equation (17.2) implies that:

$$\vec{e_1}\vec{e_1} = +1 \qquad \vec{e_2}\vec{e_2} = +1 \qquad \vec{e_3}\vec{e_3} = -1$$

$$\vec{e_1}\vec{e_2}\vec{e_1}\vec{e_2} = -1 \qquad \vec{e_1}\vec{e_3}\vec{e_1}\vec{e_3} = +1 \qquad \vec{e_2}\vec{e_3}\vec{e_2}\vec{e_3} = +1 \tag{17.4}$$

$$\vec{e_1}\vec{e_2}\vec{e_3}\vec{e_1}\vec{e_2}\vec{e_3} = +1$$

Thus we might take the subscripts, $i \& j$, to indicate the numbers of basis vectors which are square roots of plus unity and the numbers of basis vectors which are square roots of minus unity. Note that this is not a complete list of the roots of plus or minus unity because, as shown in (17.4), the basis bi-vectors and the basis multi-vectors in general are also square roots of plus or minus unity.

All traditional Clifford algebras become division algebras when we take the exponential, the polar form, of the algebra. We do this most conveniently by writing the traditional Clifford algebra as a matrix of orthogonal permutation matrices, each of which is multiplied by a variable, and taking the matrix exponential. In their division algebra form, there is no separation of elements of the algebra into vectors, bi-vectors, or multi-vectors, and the whole concept of vectors and multi-vectors is dispensed with in the division algebra form of the Clifford algebras. Instead of basis vectors and basis multi-vectors, we have imaginary units. In its division algebra form, an 8-dimensional Clifford algebra has one real variable, the scalar, and seven imaginary variables which can be thought of as either square roots of minus unity or square roots of plus unity.

We have the 8-dimensional Clifford algebras:

$$Cl_{3,0} \simeq Cl_{1,2} = \exp\left(1 + 3\sqrt{+1} + 4\sqrt{-1}\right)_{Non-Com}$$

$$Cl_{0,3} = \exp\left(1 + \sqrt{+1} + 6\sqrt{-1}\right)_{Non-Com} \tag{17.5}$$

$$Cl_{2,1} = \exp\left(1 + 5\sqrt{+1} + 2\sqrt{-1}\right)_{Non-Com}$$

In practice, we omit the 'exp' and the bracket in our denotation of the Clifford algebras, and we sometimes use commas instead of plus signs. We even omit the '1' sometimes.

First, the commutative 8-dimensional Clifford algebras:

Of the 8 sets of one hundred and twenty-eight 8-dimensional Clifford algebras, one set is comprised of only commutative algebras. Other than a few brief mentions, we have no interest in these commutative algebras. Traditionally, the commutative Clifford algebras are not considered to be Clifford algebras.

The 128 commutative 8-dimensional Clifford algebras are of two algebraically isomorphic types; these algebras are:

The Commutative 8-dimensional Clifford Algebras		Sub-algebras					
		Com		Non-com		Com	
Number of isomorphic algebras	Algebra	$\mathbb{C}$	$\mathbb{S}$	$\mathbb{H}$	A_3	A_2	A_1
16	$\exp\left(1+7\sqrt{+1}\right)_{Com}$	0	7	0	0	0	7
112	$\exp\left(1+3\sqrt{+1}+4\sqrt{-1}\right)_{Com}$	4	3	0	0	6	1

We have given the type of sub-groups; for example, the $\exp\left(1+7\sqrt{+1}_{Com}\right)$ algebras all have zero copies of the 2-dimensional complex numbers, $\mathbb{C}$, seven copies of the 2-dimensional hyperbolic complex numbers, $\mathbb{S}$, and seven copies of the 4-dimensional A_1 algebras as sub-algebras. Note that the four 4-dimensional division algebras, $A_1, A_2, A_3, \mathbb{H}$, are the only division algebras derived from the commutative $C_2 \times C_2$ group. The A_1 & A_2 algebras are commutative. The A_3 & $\mathbb{H}$ algebras are non-commutative.

The non-commutative Clifford algebras:

There are seven sets of 128 non-commutative Clifford algebras. Each of the seven sets contains the same set of algebras, and so we will be concerned with only one set.

One such set of 128 non-commutative Clifford algebras is:

The Non-Commutative Clifford Algebras								
			Com		Non-com		Com	
No. of isomorphic algebras	Algebra	Algebra	$\mathbb{C}$	$\mathbb{S}$	$\mathbb{H}$	A_3	A_2	A_1
16	$1+\sqrt{+1}+6\sqrt{-1}$	$Cl_{0,3}$	6	1	4	0	3	0
64	$3\sqrt{+1}+4\sqrt{-1}_{Non-Com}$	$Cl_{3,0}$	4	3	1	3	3	0
		$Cl_{1,2}$	4	3	1	3	3	0
48	$5\sqrt{+1}+2\sqrt{-1}_{Non-Com}$	$Cl_{2,1}$	2	5	0	4	1	2

Sub-algebras of the non-commutative Clifford algebras:

Every 8-dimensional Clifford algebra has seven 2-dimensional sub-algebras and seven 4-dimensional sub-algebras. This is because the $C_2 \times C_2 \times C_2$ group has seven order two sub-groups and seven order four sub-groups. The seven order four sub-groups are all copies of the $C_2 \times C_2$ group.

The 2-dimensional sub-algebras are of the form of either the complex numbers, $\mathbb{C}$, or the hyperbolic complex numbers, $\mathbb{S}$. These are the only 2-dimensional division algebras; they derive from the finite group C_2, and each algebra corresponds to one of the seven order 2 sub-groups of the $C_2 \times C_2 \times C_2$ finite group. Each 2-dimensional sub-algebra is comprised of the real variable and one of the seven imaginary variables.

The 4-dimensional sub-algebras are either the non-commutative quaternions, $\mathbb{H}$, the non-commutative A_3 algebra, the commutative A_2 algebra, or the commutative A_1 algebra. These are the only 4-dimensional division algebras that derive from the finite group $C_2 \times C_2$ and each algebra corresponds to one of the seven order 4 sub-groups of the $C_2 \times C_2 \times C_2$ finite group. Both the right-chiral forms and the left-chiral forms of the non-commutative 4-dimensional algebras appear.

There are other 4-dimensional division algebras which derive from the C_4 finite group, but these division algebras do not appear as sub-algebras of the 8-dimensional Clifford algebras because C_4 is not a sub-group of the $C_2 \times C_2 \times C_2$ finite group.

We see that each 8-dimensional Clifford algebra has four non-commutative 4-dimensional sub-algebras and three commutative 4-dimensional sub-algebras. All the 2-dimensional sub-algebras are commutative, of course.

8-dimensional 4-dimensional sub-algebras:

Although we refer to the sub-algebras as being 2-dimensional when they are comprised of two elements of the Clifford algebra and as being 4-dimensional when they are comprised of four elements of the Clifford algebra, these sub-algebras are not 'really' 2-dimensional or 4-dimensional and they are certainly not the same as the actual 2-dimensional algebras or the actual 4-dimensional algebras. It would be more accurate, if a little strange sounding, to refer to these sub-algebras as 8-dimensional 2-dimensional sub-algebras and as 8-dimensional 4-dimensional sub-algebras respectively. An example from the quaternions will demonstrate this; we have the quaternion and the complex number:

$$\mathbb{H} = \begin{bmatrix} a & b & c & d \\ -b & a & -d & c \\ -c & d & a & -b \\ -d & -c & b & a \end{bmatrix} \qquad \& \qquad \mathbb{C} = \begin{bmatrix} a & b \\ -b & a \end{bmatrix} \tag{17.6}$$

The quaternions have three 2-dimensional sub-algebras which are of the same nature as the complex numbers, $\mathbb{C}$. We choose one 2-dimensional sub-algebra and compare it with the 2-dimensional complex numbers. We take the polar forms to assist our exposition:

$$\exp\left(\begin{bmatrix} a & b & 0 & 0 \\ -b & a & 0 & 0 \\ 0 & 0 & a & -b \\ 0 & 0 & b & a \end{bmatrix}\right) = e^a \begin{bmatrix} \cos b & \sin b & 0 & 0 \\ -\sin b & \cos b & 0 & 0 \\ 0 & 0 & \cos b & -\sin b \\ 0 & 0 & \sin b & \cos b \end{bmatrix} \neq e^a \begin{bmatrix} \cos b & \sin b & 0 & 0 \\ -\sin b & \cos b & 0 & 0 \\ 0 & 0 & 1 & 0 \\ 0 & 0 & 0 & 1 \end{bmatrix} \tag{17.7}$$

$$\exp\left(\begin{bmatrix} a & b \\ -b & a \end{bmatrix}\right) = e^a \begin{bmatrix} \cos b & \sin b \\ -\sin b & \cos b \end{bmatrix}$$

The quaternion 2-dimensional sub-algebra is a double cover of the 2-dimensional complex numbers. We have rotation in both the clockwise direction and the anti-clockwise direction in the quaternion sub-algebra but rotation in only one direction in the complex numbers. A complete rotation in the complex numbers takes 360^0. A complete rotation in the quaternion sub-algebra takes 720^0.

Taking the eigenvectors of the quaternion rotation matrix, (17.7), will show that this '2-dimensional' rotation is not 2-dimensional rotation about an axis, or two axes in a 4-dimensional space; this is still a 4-dimensional rotation. The angle in this 2-dimensional 4-dimensional quaternion sub-algebra is still a 4-dimensional angle. This contrasts with the angle in the 2-dimensional complex numbers which is a 2-dimensional angle, of course.

In short, we must be wary about the dimensionality of the sub-algebras.

Chirality:

Non-commutative algebras in which $BC = -CB$ and either the commutator or the anti-commutator is zero for any pair of variables have chirality. The 4-dimensional Clifford algebras have chirality. The 8-dimensional Clifford algebras have chirality. The 16-dimensional Clifford algebras appear to have chirality, but notation can be deceptive; when we put the 16-dimensional Clifford algebras into matrix format, we will see that they do not have chirality.

Within the 4-dimensional Clifford algebras, the chirality of an algebra is manifest as the commutation relations of the algebra. The commutation relations are a consequence of the distribution of minus signs within the algebraic matrix form of the algebra. All of this still applies within the 8-dimensional Clifford algebras, but things are more complicated in the 8-dimensional Clifford algebras.

Different algebras have different chirality. The A_3 algebras have chirality equivalent to what we normally call parity. The quaternion algebras have a chirality which we normally call CP invariance, and they do not respect parity. The 8-dimensional algebras have a different chirality, and we must not expect them to respect either parity or CP invariance.

The algebraic chirality reference frame:

The arbitrary algebraic reference frame we use for the 8-dimensional algebras is the same as we used in the 4-dimensional algebras. We keep the whole of the 4-dimensional reference frame. The 4-dimensional reference frame is:

a) The variables in the algebraic matrix form are placed in the top row of the algebraic matrix form in alphabetic order from left to right.
b) There will be no minus signs on the top row of the algebraic matrix form.
c) We will consider the absolute values of the variables when deciding chirality.
d) The order of multiplication of two variables is standard matrix multiplication from left to right.
e) The order of the variables being multiplied will be alphabetic from left to right.

Differentials:

The differentials of the 8-dimensional Clifford algebras are formed analogously to the differentials of the 4-dimensional Clifford algebras. The potentials are written as matrices, Φ, and these matrices are operated upon by the appropriate differential operator, d, to produce two differentials called the E-field and the B-field:

$$E = \frac{1}{2}(d\Phi + \Phi d)$$

$$B = \frac{1}{2}(d\Phi - \Phi d)$$

(17.8)

The General 8-dimensional Matrix Form

We use the matrix representation of the 8-dimensional Clifford algebras. This is necessarily the matrix representation of the $C_2 \times C_2 \times C_2$ finite group. The set of all 8×8 permutation matrices is the symmetric finite group S_8 of order 40,320; this group contains as sub-groups many copies of the finite group $C_2 \times C_2 \times C_2$[100]. We will work with only the copy given below. We assume that what we shall discover about this copy of $C_2 \times C_2 \times C_2$ is everything there is to discover about the 8-dimensional Clifford algebras.

The Cayley table:

There is usually more than one Standard Form of Cayley table for a particular finite group[101]. Throughout this book, we will be using the particular Standard Form Cayley table of the $C_2 \times C_2 \times C_2$ finite group presented below, (18.1).

The Standard Form Cayley table of the $C_2 \times C_2 \times C_2$ finite group which we shall use is:

$$C_2 \times C_2 \times C_2 = \begin{bmatrix} a & b & c & d & e & f & g & h \\ b & a & d & c & f & e & h & g \\ c & d & a & b & g & h & e & f \\ d & c & b & a & h & g & f & e \\ e & f & g & h & a & b & c & d \\ f & e & h & g & b & a & d & c \\ g & h & e & f & c & d & a & b \\ h & g & f & e & d & c & b & a \end{bmatrix} \tag{18.1}$$

Within this form of Cayley table, each variable is a permutation matrix multiplied by a real number. Thus each separate variable matrix is an element of the group.

Within this form, both the a variable and the e variable commute with all other variables; in other words, the centre of the group $C_2 \times C_2 \times C_2$ is of order two being the identity and one other variable; in this case, the centre is $\{a, e\}$.

---

[100] Your author has no idea how many $C_2 \times C_2 \times C_2$ subgroups are within S_8.

[101] Each different Standard Form Cayley table corresponds to a different sub-group of the symmetric group of all permutation matrices of the appropriate size. There are as many different Standard Form Cayley tables of a given size, n, as there are subgroups of the symmetric group of order $n!$. Some of these order n sub-groups might be the same type of group.

The general matrix form of the C₂ x C₂ x C₂ algebras:

There are eight general algebraic matrix forms of the $C_2 \times C_2 \times C_2$ algebras based upon the Standard Form Cayley table above, (18.1). Of these eight general matrix forms of the $C_2 \times C_2 \times C_2$ algebras, one form is commutative and holds 128 commutative division algebras and the other seven forms are non-commutative and each hold 128 non-commutative division algebras.

The commutative 8-dimensional algebras:

The general algebraic matrix form of the $C_2 \times C_2 \times C_2$ commutative algebras is a 8×8 matrix, (18.2):

Commutative

$$
\begin{bmatrix}
a & b & c & d & e & f & g & h \\[6pt]
P_{2,1}b & a & \frac{P_{2,1}}{P_{2,4}}d & P_{2,4}c & \frac{P_{2,1}}{P_{2,6}}f & P_{2,6}e & \frac{P_{2,1}}{P_{2,8}}h & P_{2,8}g \\[6pt]
P_{3,1}c & \frac{P_{3,1}}{P_{2,4}}d & a & P_{2,4}b & \frac{P_{3,1}}{P_{3,7}}g & \frac{P_{2,6}P_{3,1}}{P_{2,8}P_{3,7}}h & P_{3,7}e & \frac{P_{2,8}P_{3,7}}{P_{2,6}}f \\[6pt]
\frac{P_{2,1}P_{3,1}}{P_{2,4}^2}d & \frac{P_{3,1}}{P_{2,4}}c & \frac{P_{2,1}}{P_{2,4}}b & a & \frac{P_{2,1}P_{3,1}}{P_{2,4}P_{2,8}P_{3,7}}h & \frac{P_{2,6}P_{3,1}}{P_{2,4}P_{3,7}}g & \frac{P_{2,1}P_{3,7}}{P_{2,4}P_{2,6}}f & \frac{P_{2,8}P_{3,7}}{P_{2,4}}e \\[6pt]
P_{5,1}e & \frac{P_{5,1}}{P_{2,6}}f & \frac{P_{5,1}}{P_{3,7}}g & \frac{P_{2,4}P_{5,1}}{P_{2,8}P_{3,7}}h & a & P_{2,6}b & P_{3,7}c & \frac{P_{2,8}P_{3,7}}{P_{2,4}}d \\[6pt]
\frac{P_{2,1}P_{5,1}}{P_{2,6}^2}f & \frac{P_{5,1}}{P_{2,6}}e & \frac{P_{2,1}P_{5,1}}{P_{2,6}P_{2,8}P_{3,7}}h & \frac{P_{2,4}P_{5,1}}{P_{2,6}P_{3,7}}g & \frac{P_{2,1}}{P_{2,6}}b & a & \frac{P_{2,1}P_{3,7}}{P_{2,4}P_{2,6}}d & \frac{P_{2,8}P_{3,7}}{P_{2,6}}c \\[6pt]
\frac{P_{3,1}P_{5,1}}{P_{3,7}^2}g & \frac{P_{3,1}P_{5,1}}{P_{2,8}P_{3,7}^2}h & \frac{P_{5,1}}{P_{3,7}}e & \frac{P_{2,4}P_{5,1}}{P_{2,6}P_{3,7}}f & \frac{P_{3,1}}{P_{3,7}}c & \frac{P_{2,6}P_{3,1}}{P_{2,4}P_{3,7}}d & a & P_{2,8}b \\[6pt]
\frac{P_{2,1}P_{3,1}P_{5,1}}{P_{2,8}^2P_{3,7}^2}h & \frac{P_{3,1}P_{5,1}}{P_{2,8}P_{3,7}^2}g & \frac{P_{2,1}P_{5,1}}{P_{2,6}P_{2,8}P_{3,7}}f & \frac{P_{2,4}P_{5,1}}{P_{2,8}P_{3,7}}e & \frac{P_{2,1}P_{3,1}}{P_{2,4}P_{2,8}P_{3,7}}d & \frac{P_{2,6}P_{3,1}}{P_{2,8}P_{3,7}}c & \frac{P_{2,1}}{P_{2,8}}b & a
\end{bmatrix}
$$

(18.2)

There are seven scaling parameters, $\{P_{2,1}, P_{2,4}, P_{2,6}, P_{2,8}, P_{3,1}, P_{3,7}, P_{5,1}\}$. Setting the various scaling parameters, $P_{i,j}$, equal to the various permutations of ± 1 in the above matrix, (18.2), will give $2^7 = 128$ commutative algebras.

Each of these algebras is formed by taking the exponential of the matrix with a particular permutation of the parameters. In general, the matrix alone is not a division algebra; it is only the exponential form that satisfies the requirements necessary to be a division algebra.

These 128 commutative division algebras are of two distinct, algebraically non-isomorphic, types. There are sixteen $1 + 7\sqrt{+1}$ commutative algebras. There are one hundred and twelve $1 + 3\sqrt{+1} + 4\sqrt{-1}_{Com}$ commutative algebras. Note that these algebras do not actually include the $\sqrt{+1}$ elements; the exponentiation rids the matrix form of these elements. These elements appear as asymptotically approached parts of the algebra - they are just outside of the algebra by infinitesimal amounts.

These commutative algebras are sister algebras of the 8-dimensional Clifford algebras, but, being commutative, they are not usually considered to be Clifford algebras.

The non-commutative C_2 x C_2 x C_2 algebras:

There are seven general algebraic matrix forms which each produce 128 non-commutative $C_2 \times C_2 \times C_2$ algebras (8-dimensional Clifford algebras). Each of these seven forms produces the same set of 64 non-commutative $1+3\sqrt{+1}+4\sqrt{-1}_{Non-Com}$ algebras, 48 non-commutative $1+5\sqrt{+1}+2\sqrt{-1}$ algebras, and 16 non-commutative $1+1\sqrt{+1}+6\sqrt{-1}$ algebras. The seven sets differ by the permutations of the three parameters, $\{S_4, S_6, S_7\} = \pm 1$ whose distribution is given in the matrix below, (18.3). The permutation with all three of these parameters equal to plus unity gives the 8-dimensional commutative $C_2 \times C_2 \times C_2$ 'sister' algebras given above.

The matrix of permutations of the three parameters, $\{S_4, S_6, S_7\} = \pm 1$ is:

$$
\begin{bmatrix}
a & b & c & d & e & f & g & h \\
\sim & a & \sim & \sim & \sim & \sim & \sim & \sim \\
\sim & S_4 & a & S_4 & \sim & S_4 & \sim & S_4 \\
S_4 & \sim & S_4 & a & S_4 & \sim & S_4 & \sim \\
\sim & S_6 S_7 & S_7 & S_6 & a & S_6 S_7 & S_7 & S_6 \\
S_6 S_7 & \sim & S_6 & S_7 & S_6 S_7 & a & S_6 & S_7 \\
S_7 & S_4 S_6 & \sim & S_4 S_6 S_7 & S_7 & S_4 S_6 & a & S_4 S_6 S_7 \\
S_4 S_6 & S_7 & S_4 S_6 S_7 & \sim & S_4 S_6 & S_7 & S_4 S_6 S_7 & a
\end{bmatrix}
\tag{18.3}
$$

Since the seven sets of non-commutative algebras are all the same, we will study only one set of 64 non-commutative $1+3\sqrt{+1}+4\sqrt{-1}_{Non-Com}$ algebras, 48 non-commutative $1+5\sqrt{+1}+2\sqrt{-1}$ algebras, and 16 non-commutative $1+1\sqrt{+1}+6\sqrt{-1}$ algebras.

The general matrix form of the $C_2 \times C_2 \times C_2$ non-commutative algebras is an 8×8 matrix. Note that the distribution of minus signs is based upon $S_4 = -1$, $S_6 = S_7 = +1$; this gives:

$$\begin{bmatrix} a & b & c & d & e & f & g & h \\ b & a & \sim & \sim & \sim & \sim & \sim & \sim \\ c & -1 & a & -1 & \sim & -1 & \sim & -1 \\ -1 & \sim & -1 & a & -1 & \sim & -1 & \sim \\ e & +1 & +1 & +1 & a & +1 & +1 & +1 \\ +1 & \sim & +1 & +1 & +1 & a & +1 & +1 \\ +1 & -1 & \sim & -1 & +1 & -1 & a & -1 \\ -1 & +1 & -1 & \sim & -1 & +1 & -1 & a \end{bmatrix}$$

(18.4)

The general non-commutative matrix form:

A general algebraic matrix form for the non-commutative 8-dimensional $C_2 \times C_2 \times C_2$ algebras is:

$$Non-Commutative$$

$$\begin{bmatrix} a & b & c & d & e & f & g & h \\[4pt] P_{2.1}b & a & \dfrac{P_{2.1}}{P_{2.4}}d & P_{2.4}c & \dfrac{P_{2.1}}{P_{2.6}}f & P_{2.6}e & \dfrac{P_{2.1}}{P_{2.8}}h & P_{2.8}g \\[8pt] P_{3.1}c & -\dfrac{P_{3.1}}{P_{2.4}}d & a & -P_{2.4}b & \dfrac{P_{3.1}}{P_{3.7}}g & -\dfrac{P_{2.6}P_{3.1}}{P_{2.8}P_{3.7}}h & P_{3.7}e & -\dfrac{P_{2.8}P_{3.7}}{P_{2.6}}f \\[8pt] -\dfrac{P_{2.1}P_{3.1}}{P_{2.4}^2}d & \dfrac{P_{3.1}}{P_{2.4}}c & -\dfrac{P_{2.1}}{P_{2.4}}b & a & -\dfrac{P_{2.1}P_{3.1}}{P_{2.4}P_{2.8}P_{3.7}}h & \dfrac{P_{2.6}P_{3.1}}{P_{2.4}P_{3.7}}g & -\dfrac{P_{2.1}P_{3.7}}{P_{2.4}P_{2.6}}f & \dfrac{P_{2.8}P_{3.7}}{P_{2.4}}e \\[8pt] P_{5.1}e & \dfrac{P_{5.1}}{P_{2.6}}f & \dfrac{P_{5.1}}{P_{3.7}}g & \dfrac{P_{2.4}P_{5.1}}{P_{2.8}P_{3.7}}h & a & P_{2.6}b & P_{3.7}c & \dfrac{P_{2.8}P_{3.7}}{P_{2.4}}d \\[8pt] \dfrac{P_{2.1}P_{5.1}}{P_{2.6}^2}f & \dfrac{P_{5.1}}{P_{2.6}}e & \dfrac{P_{2.1}P_{5.1}}{P_{2.6}P_{2.8}P_{3.7}}h & \dfrac{P_{2.4}P_{5.1}}{P_{2.6}P_{3.7}}g & \dfrac{P_{2.1}}{P_{2.6}}b & a & \dfrac{P_{2.1}P_{3.7}}{P_{2.4}P_{2.6}}d & \dfrac{P_{2.8}P_{3.7}}{P_{2.6}}c \\[8pt] \dfrac{P_{3.1}P_{5.1}}{P_{3.7}^2}g & -\dfrac{P_{3.1}P_{5.1}}{P_{2.8}P_{3.7}^2}h & \dfrac{P_{5.1}}{P_{3.7}}e & -\dfrac{P_{2.4}P_{5.1}}{P_{2.6}P_{3.7}}f & \dfrac{P_{3.1}}{P_{3.7}}c & -\dfrac{P_{2.6}P_{3.1}}{P_{2.4}P_{3.7}}d & a & -P_{2.8}b \\[8pt] -\dfrac{P_{2.1}P_{3.1}P_{5.1}}{P_{2.8}^2P_{3.7}^2}h & \dfrac{P_{3.1}P_{5.1}}{P_{2.8}P_{3.7}^2}g & -\dfrac{P_{2.1}P_{5.1}}{P_{2.6}P_{2.8}P_{3.7}}f & \dfrac{P_{2.4}P_{5.1}}{P_{2.8}P_{3.7}}e & -\dfrac{P_{2.1}P_{3.1}}{P_{2.4}P_{2.8}P_{3.7}}d & \dfrac{P_{2.6}P_{3.1}}{P_{2.8}P_{3.7}}c & -\dfrac{P_{2.1}}{P_{2.8}}b & a \end{bmatrix}$$

(18.5)

Although we give the full matrix form, since the parameters, $P_{i,j}$, will be set to ± 1, we could replace the even powers of the parameters with a one thereby simplifying the matrix a little. With this in mind, the reader's attension is drawn to the leftmost column of the above matrix, (18.5); this is beautiful mathematics.

We will be examining the 128 algebras given by taking the exponential of the above matrix, (18.5) for the 128 different permutations of $P_{i,j} = \pm 1$.

The centre of the algebras:

The elements $\{a,e\}$ each commute with all other elements of the algebras. These two elements correspond to the real part of the Clifford algebra and the tri-vector part of the Clifford algebra.

We now give an example of each type of algebra.

The first division algebra:

There are forty-eight 8-dimensional algebras of the form $5\sqrt{+1}, 2\sqrt{-1}_{non-com}$; an example is:

$$5\sqrt{+1}, 2\sqrt{-1}_{non-com} =
\exp\left(\begin{bmatrix}
a & b & c & d & e & f & g & h \\
b & a & d & c & f & e & h & g \\
c & -d & a & -b & g & -h & e & -f \\
-d & c & -b & a & -h & g & -f & e \\
e & f & g & h & a & b & c & d \\
f & e & h & g & b & a & d & c \\
g & -h & e & -f & c & -d & a & -b \\
-h & g & -f & e & -d & c & -b & a
\end{bmatrix}\right) \tag{18.6}$$

The second division algebra:

There are sixty-four 8-dimensional algebras of the form $3\sqrt{+1}, 4\sqrt{-1}_{non-com}$; an example is:

$$3\sqrt{+1}, 4\sqrt{-1}_{non-com} =
\exp\left(\begin{bmatrix}
a & b & c & d & e & f & g & h \\
b & a & d & c & f & e & h & g \\
c & -d & a & -b & g & -h & e & -f \\
-d & c & -b & a & -h & g & -f & e \\
-e & -f & -g & -h & a & b & c & d \\
-f & -e & -h & -g & b & a & d & c \\
-g & h & -e & f & c & -d & a & -b \\
h & -g & f & -e & -d & c & -b & a
\end{bmatrix}\right) \tag{18.7}$$

The third division algebra:

There are sixteen 8-dimensional algebras of the form $1\sqrt{+1}, 6\sqrt{-1}_{non-com}$; an example is:

$$1\sqrt{+1}, 6\sqrt{-1}_{non-com} =$$

$$\exp\left(\begin{bmatrix} a & b & c & d & e & f & g & h \\ -b & a & -d & c & -f & e & -h & g \\ -c & d & a & -b & -g & h & e & -f \\ -d & -c & b & a & -h & -g & f & e \\ e & f & g & h & a & b & c & d \\ -f & e & -h & g & -b & a & -d & c \\ -g & h & e & -f & -c & d & a & -b \\ -h & -g & f & e & -d & -c & b & a \end{bmatrix}\right) \qquad (18.8)$$

With a little thought, we realise that, since all variables are either symmetric or anti-symmetric, and since anti-symmetric variables correspond to square roots of minus unity while symmetric variables correspond to square roots of plus unity, and since the elements of the top row have do not have minus signs, we can see the form of the algebra by simply counting the number of minus signs in the left-most column of the matrix; each minus sign corresponds to a square root of minus unity, and the absence of a minus sign corresponds to a square root of plus unity excepting the identity (the real number).

We see that there are 16 non-commutative $1+1\sqrt{+1}+6\sqrt{-1}$ algebras because there are 16 ways to choose the signs of the parameters, $P_{i,j}$, so as to produce six minus signs in the leftmost column:

$$P_{2,1} = P_{3,1} = -1, \qquad P_{5,1} = +1$$
$$P_{2,4} = \pm1, \quad P_{2,6} = \pm1, \quad P_{2,8} = \pm1, \quad P_{3,7} = \pm1 \qquad (18.9)$$

Similar considerations apply to the other algebras.

An asymmetry and a look ahead:

Given the simple symmetry in this area of mathematics, we are surprised to find the 128 algebras divided into a set of 64 alegbraically isomorphic algebras and a further set of 64 algebras which is separated into two algebraically distinct types of algebra. We might have expected two sets of 64 algebras or perhaps four sets of 32 algebras. We will eventually extract emergent E-fields from each of these three sets of algebras and identify these emergent E-fields with '8-dimensional electrons' which we will take to be quarks. We thus have three types of quark when, with one eye on the Standard Model of particle physics, we might have expected only two types of quarks. It will transpire that the quark from the set of 16 algebras is indistinguishable from the quark from the set of 48 algebras; thus we have only two types of quarks. It will transpire that the electrical field of the quark from the set of 64 algebras is twice the electrical field of either of the quarks from the other two sets of algebras. In other words, we observe two quarks and their relative electrical charges are 2:1; this is just what we have in the Standard Model.

Chapter 19

Commutation Relations of the 8-dimensional Algebras

Before we wade into the 8-dimensional chiral algebras, let us take a brief look at what we might find.

Violation of CP invariance:

Looking at the general non-commutative 8-dimensional algebraic matrix form, (18.5), consider the 8-dimensional algebra formed with:

$$P_{2,1} = P_{3,1} = P_{2,6} = P_{3,7} = -1, \qquad P_{2,4} = P_{2,8} = +1 \tag{19.1}$$

For clarity, we set the variables $e = f = g = h = 0$

$$\begin{bmatrix} a & b & c & d & 0 & 0 & 0 & 0 \\ -b & a & -d & c & 0 & 0 & 0 & 0 \\ -c & d & a & -b & 0 & 0 & 0 & 0 \\ -d & -c & b & a & 0 & 0 & 0 & 0 \\ 0 & 0 & 0 & 0 & a & -b & -c & -d \\ 0 & 0 & 0 & 0 & b & a & -d & c \\ 0 & 0 & 0 & 0 & c & d & a & -b \\ 0 & 0 & 0 & 0 & d & -c & b & a \end{bmatrix} \tag{19.2}$$

We see that the top left-hand 4×4 corner of the matrix, (19.2), is a left-chiral quaternion. We see that the bottom right-hand 4×4 corner of the matrix, (19.2), is a conjugate right-chiral quaternion. The quaternion chirality, which is CP invariance, is violated within the 8-dimensional algebras.

The violation of CP invariance is not universal within the 8-dimensional algebras. We have:

$$P_{2,1} = P_{3,1} = P_{2,4} = P_{2,6} = P_{3,7} = -1, \qquad P_{2,8} = +1 \tag{19.3}$$

$$\begin{bmatrix} a & b & c & d & 0 & 0 & 0 & 0 \\ -b & a & d & -c & 0 & 0 & 0 & 0 \\ -c & -d & a & b & 0 & 0 & 0 & 0 \\ -d & c & -b & a & 0 & 0 & 0 & 0 \\ 0 & 0 & 0 & 0 & a & -b & -c & d \\ 0 & 0 & 0 & 0 & b & a & d & c \\ 0 & 0 & 0 & 0 & c & -d & a & -b \\ 0 & 0 & 0 & 0 & -d & -c & b & a \end{bmatrix} \tag{19.4}$$

We see that the top left-hand 4×4 corner of the matrix, (19.4), is a right-chiral quaternion. We see that the bottom right-hand 4×4 corner of the matrix, (19.4), is also a right-chiral quaternion. The quaternion chirality, which is CP invariance, is not violated within this 8-dimensional algebra.

Algebras are married:

The centre of the $C_2 \times C_2 \times C_2$ group is the two elements $\{a,e\}$. Thus there are two elements of every 8-dimensional Clifford algebra which commute with every other element of the algebra. We therefore have commutation relations between only six elements $\{b,c,d,f,g,h\}$.

The commutation relations between the particular variables are dependent upon the general matrix form (18.5). Consider the product of two variables extracted from the general matrix form (18.5):

$$
\begin{bmatrix}
0 & b & 0 & 0 & 0 & 0 & 0 & 0 \\
P_{2,1}b & 0 & 0 & 0 & 0 & 0 & 0 & 0 \\
0 & 0 & 0 & -P_{2,4}b & 0 & 0 & 0 & 0 \\
0 & 0 & -\dfrac{P_{2,1}}{P_{2,4}}b & 0 & 0 & 0 & 0 & 0 \\
0 & 0 & 0 & 0 & 0 & P_{2,6}b & 0 & 0 \\
0 & 0 & 0 & 0 & \dfrac{P_{2,1}}{P_{2,6}}b & 0 & 0 & 0 \\
0 & 0 & 0 & 0 & 0 & 0 & 0 & -P_{2,8}b \\
0 & 0 & 0 & 0 & 0 & 0 & -\dfrac{P_{2,1}}{P_{2,8}}b & 0
\end{bmatrix}
\tag{19.5}
$$

And:

$$
\begin{bmatrix}
0 & 0 & c & 0 & 0 & 0 & 0 & 0 \\
0 & 0 & 0 & P_{2,4}c & 0 & 0 & 0 & 0 \\
P_{3,1}c & 0 & 0 & 0 & 0 & 0 & 0 & 0 \\
0 & \dfrac{P_{3,1}}{P_{2,4}}c & 0 & 0 & 0 & 0 & 0 & 0 \\
0 & 0 & 0 & 0 & 0 & 0 & P_{3,7}c & 0 \\
0 & 0 & 0 & 0 & 0 & 0 & 0 & \dfrac{P_{2,8}P_{3,7}}{P_{2,6}}c \\
0 & 0 & 0 & 0 & \dfrac{P_{3,1}}{P_{3,7}}c & 0 & 0 & 0 \\
0 & 0 & 0 & 0 & 0 & \dfrac{P_{2,6}P_{3,1}}{P_{2,8}P_{3,7}}c & 0 & 0
\end{bmatrix}
\tag{19.6}
$$

The commutator of these two matrices will always be the d variable. We have the product of (19.5) & (19.6) in alphabetic order, $\begin{bmatrix} b & c \end{bmatrix}$:

$$d = 2bc$$

$$\begin{bmatrix} 0 & 0 & 0 & P_{2,4}d & 0 & 0 & 0 & 0 \\ 0 & 0 & P_{2,1}d & 0 & 0 & 0 & 0 & 0 \\ 0 & -P_{3,1}d & 0 & 0 & 0 & 0 & 0 & 0 \\ -\dfrac{P_{2,1}P_{3,1}}{P_{2,4}}d & 0 & 0 & 0 & 0 & 0 & 0 & 0 \\ 0 & 0 & 0 & 0 & 0 & 0 & 0 & P_{2,8}P_{3,7}d \\ 0 & 0 & 0 & 0 & 0 & 0 & \dfrac{P_{2,1}P_{3,7}}{P_{2,6}}d & 0 \\ 0 & 0 & 0 & 0 & 0 & -\dfrac{P_{3,1}P_{2,6}}{P_{3,7}}d & 0 & 0 \\ 0 & 0 & 0 & 0 & -\dfrac{P_{2,1}P_{3,1}}{P_{2,8}P_{3,7}}d & 0 & 0 & 0 \end{bmatrix} \quad (19.7)$$

We see the values of the parameters do not affect the commutation relations as far as which particular variables are concerned. The product of the b variable and the c variable, in any order, will always be the d variable.

However, the signs of the parameters, $P_{i,j}$, do affect the sign of the product variable, (19.7). The signs of the product variables is the chirality of the matrices.

Let us consider the algebra given by all parameters equal to plus unity. The d variable, (19.7), becomes:

$$P_{2,1} = P_{2,4} = P_{2,6} = P_{2,8} = P_{3,1} = P_{3,7} = +1$$

$$\begin{bmatrix} 0 & 0 & 0 & d & 0 & 0 & 0 & 0 \\ 0 & 0 & d & 0 & 0 & 0 & 0 & 0 \\ 0 & -d & 0 & 0 & 0 & 0 & 0 & 0 \\ -d & 0 & 0 & 0 & 0 & 0 & 0 & 0 \\ 0 & 0 & 0 & 0 & 0 & 0 & 0 & d \\ 0 & 0 & 0 & 0 & 0 & 0 & d & 0 \\ 0 & 0 & 0 & 0 & 0 & -d & 0 & 0 \\ 0 & 0 & 0 & 0 & -d & 0 & 0 & 0 \end{bmatrix} \quad (19.8)$$

Since the variable on the top row is positive, we might call this a left-chiral algebra. Now consider a different algebra:

$$P_{2,1} = P_{2,4} = P_{3,1} = +1 \qquad P_{2,6} = P_{2,8} = P_{3,7} = -1$$

$$\begin{bmatrix} 0 & 0 & 0 & d & 0 & 0 & 0 & 0 \\ 0 & 0 & d & 0 & 0 & 0 & 0 & 0 \\ 0 & -d & 0 & 0 & 0 & 0 & 0 & 0 \\ -d & 0 & 0 & 0 & 0 & 0 & 0 & 0 \\ 0 & 0 & 0 & 0 & 0 & 0 & 0 & d \\ 0 & 0 & 0 & 0 & 0 & 0 & d & 0 \\ 0 & 0 & 0 & 0 & 0 & -d & 0 & 0 \\ 0 & 0 & 0 & 0 & -d & 0 & 0 & 0 \end{bmatrix} \qquad (19.9)$$

We see that these two different algebras have given exactly the same distribution of minus signs on the d variable. We have two different algebras with the same commutation relation in respect of this variable; by exhaustive calculation, we will see that, within the 8-dimensional Clifford algebras, we always have two algebras, $A \& B$, with identical commutation relations. Further, opposite to these two algebras, we always have two other algebras, $C \& D$, which also have identical commutation relations which are opposite to the commutation relations of $A \& B$ – we have two different algebras with exactly the same left-chirality and two different algebras with exactly the same right-chirality. Taken together, this set of four algebras form a marriage[102] of four algebras.

In the 4-dimensional algebras, the reference frame we established against which to determine the chirality of an algebra relied upon the sign of the, in alphabetic order, product variable on the top row. Looking above, (19.7), we see that the sign of the product variable on the top row is unaffected by swapping the sign of the parameters $P_{2,6}$, & $P_{2,8}$, & $P_{3,7}$.

The algebraic reference frame is insufficient to separate the algebras into individual algebras, but it is sufficient to separate them into pairs of algebras.

We see here that there are two separate algebras, with different signs of parameters, which are effectively the same algebra. The chiralities of these two separate algebras are identical. If we were to identify one of these algebras with, say, a left-chiral quark field, then, automatically, we would have two separate but identical left-chiral quark fields.

The $P_{5,1}$ parameter:

The astute reader might have noticed that we have ignored the $P_{5,1}$ parameter in this chapter. The last sentence of the next chapter will explain why we could ignore the $P_{5,1}$ parameter in this chapter.

---

[102] This is not a technical term to be found in a mathematics dictionary. Your author has just invented the term.

The One Hundred and Twenty-eight 8-dimensional Algebras

We are concerned with a single set of the seven sets of 128 non-commutative 8-dimensional Clifford algebras. These 128 separate algebras are generated by the 128 different permutations of the seven parameters, $P_{i,j} = \pm 1$, in the non-commutative algebraic matrix form given above, (18.5).

We need to be able to separately identify each algebra. We give each of the 128 algebras a case number.

We list each algebra and its corresponding parameter values. The parameter values are listed when they are equal to minus unity. The blank spaces indicate plus unity. The case numbers are the orders in which the algebras are produced using the following computer code written in Maple 17:

```
>  for P[2, 1] from -1 to 1 by 2 do
>  for P[2, 4] from -1 to 1 by 2 do
>  for P[2, 6] from -1 to 1 by 2 do
>  for P[2, 8] from -1 to 1 by 2 do
>  for P[3, 1] from -1 to 1 by 2 do
>  for P[3, 7] from -1 to 1 by 2 do
>  for P[5, 1] from -1 to 1 by 2 do
```

We will refer to the individual algebras by the case numbers listed in this chapter.

$Cl_{0,3} = 1 + 1\sqrt{+1} + 6\sqrt{-1}$: 16 Algebras							
Case Number	$P_{2,1}$	$P_{2,4}$	$P_{2,6}$	$P_{2,8}$	$P_{3,1}$	$P_{3,7}$	$P_{5,1}$
2	−1	−1	−1	−1	−1	−1	
4	−1	−1	−1	−1	−1		
10	−1	−1	−1		−1	−1	
12	−1	−1	−1		−1		
18	−1	−1		−1	−1	−1	
20	−1	−1		−1	−1		
26	−1	−1			−1	−1	
28	−1	−1			−1		
34	−1		−1	−1	−1	−1	
36	−1		−1	−1	−1		
42	−1		−1		−1	−1	
44	−1		−1		−1		
50	−1			−1	−1	−1	
52	−1			−1	−1		
58	−1				−1	−1	
60	−1				−1		

$Cl_{2,1} = 1 + 5\sqrt{+1} + 2\sqrt{-1}$: 48 Algebras							
Case Number	$P_{2,1}$	$P_{2,4}$	$P_{2,6}$	$P_{2,8}$	$P_{3,1}$	$P_{3,7}$	$P_{5,1}$
6	−1	−1	−1	−1		−1	
8	−1	−1	−1	−1			
14	−1	−1	−1				
16	−1	−1	−1				
22	−1	−1		−1		−1	
24	−1	−1		−1			
30	−1	−1				−1	
32	−1	−1					
38	−1		−1	−1		−1	
40	−1		−1	−1			
46	−1		−1			−1	
48	−1		−1	1			
54	−1			−1		−1	
56	−1			−1			
62	−1					−1	
64	−1						
66		−1	−1	−1	−1	−1	
68		−1	−1	−1	−1		
70		−1	−1	−1		−1	
72		−1	−1	−1			
74		−1	−1		−1	−1	
76		−1	−1		−1		
78		−1	−1			−1	
80		−1	−1				
82		−1		−1	−1	−1	
84		−1		−1	−1		
86		−1		−1		−1	
88		−1		−1			
90		1			−1	−1	
92		−1			−1		
94		−1				−1	
96		−1					
98			−1	−1	−1	−1	
100			−1	−1	−1		
102			−1	−1		−1	
104			−1	−1			
106			−1		−1	−1	
108			−1		−1		
110			−1			−1	
112			−1				
114				−1	−1	−1	

116				−1	−1		
118				−1		−1	
120				−1			
122					−1	−1	
124					−1		
126						−1	
128							

$$Cl_{3,0} \cong Cl_{1,2} = 1 + 3\sqrt{+1} + 4\sqrt{-1}\,_{Non-Com} \quad : \quad \text{64 Algebras}$$

Case Number	$P_{2,1}$	$P_{2,4}$	$P_{2,6}$	$P_{2,8}$	$P_{3,1}$	$P_{3,7}$	$P_{5,1}$
1	−1	−1	−1	−1	−1	−1	−1
3	−1	−1	−1	−1	−1		−1
5	−1	−1	−1	−1		−1	−1
7	−1	−1	−1	−1			−1
9	−1	−1	−1		−1	−1	−1
11	−1	−1	−1		−1		−1
13	−1	−1	−1			−1	−1
15	−1	−1	−1				−1
17	−1	−1		−1	−1	−1	−1
19	−1	−1		−1	−1		−1
21	−1	−1		−1		−1	−1
23	−1	−1		−1			−1
25	−1	−1			−1	−1	−1
27	−1	−1			−1		−1
29	−1	−1				−1	−1
31	−1	−1					−1
33	−1		−1	−1	−1	−1	−1
35	−1		−1	−1		−1	−1
37	−1		+1	−1	−1	−1	−1
39	−1		−1	−1			−1
41	−1		−1		−1	−1	−1
43	−1		−1		−1		−1
45	−1		−1			−1	−1
47	−1		−1				−1
49	−1			−1	−1	−1	−1
51	−1			−1	−1		−1
53	−1			−1		−1	−1
55	−1			−1			−1
57	−1				−1	−1	−1
59	−1				−1		−1
61	−1					−1	−1
63	−1						−1
65		−1	−1	−1	−1	−1	−1
67		−1	−1	−1	−1		−1

69		−1	−1	−1		−1	−1
71		−1	−1	−1			−1
73		−1	−1		−1	−1	−1
75		−1	−1		−1		−1
77		−1	−1			−1	−1
79		−1	−1				−1
81		−1		−1	−1	−1	−1
83		−1		−1	−1		−1
85		−1		−1		−1	−1
87		−1		−1			−1
89		−1			−1	−1	−1
91		−1			−1		−1
93		−1				−1	−1
95		−1					−1
97			−1	−1	−1	−1	−1
99			−1	−1	−1		−1
101			−1	−1		−1	−1
103			−1	−1			−1
105					−1	−1	−1
107			−1		−1		−1
109			−1			−1	−1
111			−1				−1
113				−1	−1	−1	−1
115				−1	−1		−1
117				−1		−1	−1
119				−1			−1
121					−1	−1	−1
123					−1		−1
125						−1	−1
127							−1

We see that the sixteen $Cl_{0,3} = 1 + 1\sqrt{+1} + 6\sqrt{-1}$ algebras and the forty-eight $Cl_{2,1} = 1 + 5\sqrt{+1} + 2\sqrt{-1}$ algebras all have $P_{5,1} = +1$ and the sixty-four $Cl_{3,0} \simeq Cl_{1,2} = 1 + 3\sqrt{+1} + 4\sqrt{-1}_{Non-Com}$ algebras all have $P_{5,1} = -1$. We see that the role of the $P_{5,1}$ parameter is to determine the type of algebra.

The Full Sets of the Commutation Relations

The commutation relations of the sixteen $Cl_{0,3}$ algebras 1:

The commutation relations of the sixteen cases of the $Cl_{0,3} = 1 + 6\sqrt[2]{-1} + \sqrt[2]{+1}$ algebras are listed below. The left-most column is the case number. We see that, in terms of permutation relations, the sixteen algebras occur in eight pairs. The two members of each pair have the same commutation relations. Using the case numbers to identify the algebras, those pairings are:

$$\{2,20\}, \quad \{4,18\}, \quad \{10,28\}, \quad \{12,26\}$$
$$\{42,60\}, \quad \{44,58\}, \quad \{34,52\}, \quad \{36,50\}$$

(21.1)

Of these pairings, we have that the commutation relations of each of the $\{2,20\}$ algebras are the exact opposite to the commutation relations of each of the $\{42,60\}$ algebras; the same is true of the two pairs $\{4,18\}$ and $\{44,58\}$; the same is true of the two pairs $\{10,28\}$ and $\{34,52\}$; the same is true of the two pairs $\{12,26\}$ and $\{36,50\}$. It seems that the algebras in each identical pair, for example, $\{2,20\}$, are of the same chirality and that this chirality is opposite to the chirality of each of the pair of exactly opposite algebras, for example, $\{42,60\}$.

If two algebraically isomorphic algebras have exactly the same commutation relations, are they the same algebra? Experience in this area of mathematics inclines your author to say no, but the question is open.

The commutation relations of the sixteen $Cl_{0,3}$ algebras 2:

The relations in the below table are based upon the commutator in alphabetic order. The size of the table leads us to use the following type of abbreviation for the commutator:

$$[b \quad c] = bc - cb \quad \text{is written as} \quad bc$$

(21.2)

and other variable commutators similarly so; we ignore the 2's[103].

The a variable and the e variable commute with all other variables, and so we omit these commutation relations from the table.

When multiplied together, some variables have as their product the commutative e variable. In such cases, we have:

$$[b \quad f] = [c \quad g] = [d \quad h] = e - e = 0$$

(21.3)

---

[103] Twos are two a penny.

Of course, these commutation relations are within every algebra. In Clifford algebra terms, the e variable is the tri-vector $\overrightarrow{e_{123}}$. We see that the pairs of variables above, (21.3), like $b \& f$ are pairings of a vector and a bi-vector like $\overrightarrow{e_1}\overrightarrow{e_{23}} = \overrightarrow{e_{123}} = \overrightarrow{e_{23}}\overrightarrow{e_1}$.

Commutation relations of the sixteen $Cl_{0,3} = 1 + 6\sqrt[2]{-1} + \sqrt[2]{+1}$ algebras														
Case	bc	bd	bf	bg	bh	cd	cf	cg	ch	df	dg	dh	fg	fh	gh
2	$-d$	$+c$	0	$-h$	$+g$	$-b$	$+h$	0	$-f$	$-g$	$+f$	0	$-d$	$+c$	$-b$
4	$-d$	$+c$	0	$-h$	$+g$	$-b$	$-h$	0	$+f$	$+g$	$-f$	0	$+d$	$-c$	$-b$
10	$-d$	$+c$	0	$+h$	$-g$	$-b$	$-h$	0	$+f$	$-g$	$+f$	0	$-d$	$-c$	$+b$
12	$-d$	$+c$	0	$+h$	$-g$	$-b$	$+h$	0	$-f$	$+g$	$-f$	0	$+d$	$+c$	$+b$
18	$-d$	$+c$	0	$-h$	$+g$	$-b$	$-h$	0	$+f$	$+g$	$-f$	0	$+d$	$-c$	$-b$
20	$-d$	$+c$	0	$-h$	$+g$	$-b$	$+h$	0	$-f$	$-g$	$+f$	0	$-d$	$+c$	$-b$
26	$-d$	$+c$	0	$+h$	$-g$	$-b$	$+h$	0	$-f$	$+g$	$-f$	0	$+d$	$+c$	$+b$
28	$-d$	$+c$	0	$+h$	$-g$	$-b$	$-h$	0	$+f$	$-g$	$+f$	0	$-d$	$-c$	$+b$
34	$+d$	$-c$	0	$-h$	$+g$	$+b$	$+h$	0	$-f$	$+g$	$-f$	0	$+d$	$+c$	$-b$
36	$+d$	$-c$	0	$-h$	$+g$	$+b$	$-h$	0	$+f$	$-g$	$+f$	0	$-d$	$-c$	$-b$
42	$+d$	$-c$	0	$+h$	$-g$	$+b$	$-h$	0	$+f$	$+g$	$-f$	0	$+d$	$-c$	$+b$
44	$+d$	$-c$	0	$+h$	$-g$	$+b$	$+h$	0	$-f$	$-g$	$+f$	0	$-d$	$+c$	$+b$
50	$+d$	$-c$	0	$-h$	$+g$	$+b$	$-h$	0	$+f$	$-g$	$+f$	0	$-d$		$-b$
52	$+d$	$-c$	0	$-h$	$+g$	$+b$	$+h$	0	$-f$	$+g$	$-f$	0	$+d$	$+c$	$-b$
58	$+d$	$-c$	0	$+h$	$-g$	$+b$	$+h$	0	$-f$	$-g$	$+f$	0	$-d$	$+c$	$+b$
60	$+d$	$-c$	0	$+h$	$-g$	$+b$	$-h$	0	$+f$	$+g$	$-f$	0	$+d$	$-c$	$+b$

Identical chiral algebraically isomorphic algebras:

The sixteen $Cl_{0,3} = 1 + 6\sqrt[2]{-1} + \sqrt[2]{+1}$ algebras are algebraically isomorphic. However, individual algebras can be distinguished from each other by their commutation relations, but we see that, even taking account of commutation relations, we have two copies of every $Cl_{0,3} = 1 + 6\sqrt[2]{-1} + \sqrt[2]{+1}$ algebra, see (21.1).

Right chiral and left chiral algebras:

Since the two cases $\{2, 20\}$ are identical algebras with identical commutation relations, and since the two cases $\{42, 60\}$ are identical algebras with identical commutation relations, and since the commutation relations of the two cases $\{42, 60\}$ are exactly opposite to the commutation relations of the $\{2, 20\}$ cases, the reader might think that both the $\{2, 20\}$ algebras share the same chirality which is the opposite chirality to the chirality of both the $\{42, 60\}$ algebras. The situation is more complicated. Let us look at the 4-dimensional case.

If we add the two quaternion algebraic matrix forms with the same variables, we get:

$$
\begin{bmatrix} a & b & c & d \\ -b & a & -d & c \\ -c & d & a & -b \\ -d & -c & b & a \end{bmatrix} + \begin{bmatrix} a & b & c & d \\ -b & a & d & -c \\ -c & -d & a & b \\ -d & c & -b & a \end{bmatrix} = 2 \begin{bmatrix} a & b & c & d \\ -b & a & 0 & 0 \\ -c & 0 & a & 0 \\ -d & 0 & 0 & a \end{bmatrix} \tag{21.4}
$$

We have only the top row, the left-most column and the leading diagonal which are non-zero. The same is true of each pair of A_3 algebras. Within the 4-dimensional $C_2 \times C_2$ algebras, the algebras of opposite chirality have 'opposite' distributions of minus signs in their matrices excepting the top row and the left-most column; it is this 'oppositeness' of signs that is expressed in the presence of the zeros in the sum of two algebraic matrix forms of opposite chirality as in (21.4).

The two oppositely chiral quaternion algebraic matrix forms commute. Note we use different variables as well as different matrix forms:

$$
\begin{bmatrix} a & b & c & d \\ -b & a & -d & c \\ -c & d & a & -b \\ -d & -c & b & a \end{bmatrix} \begin{bmatrix} t & x & y & z \\ -x & t & z & -y \\ -y & -z & t & x \\ -z & y & -x & t \end{bmatrix} = \begin{bmatrix} t & x & y & z \\ -x & t & z & -y \\ -y & -z & t & x \\ -z & y & -x & t \end{bmatrix} \begin{bmatrix} a & b & c & d \\ -b & a & -d & c \\ -c & d & a & -b \\ -d & -c & b & a \end{bmatrix} \tag{21.5}
$$

$$
\mathbb{H}_{L\chi} \mathbb{H}_{R\chi} = \mathbb{H}_{R\chi} \mathbb{H}_{L\chi}
$$

The same is true of the three pairs of A_3 algebras. On the side, the reader will recall that the two commutative A_1 algebras, (3.6), do not commute; the significance of this, if any, is not understood.

Within the 4-dimensional Clifford algebras, we have found two properties of oppositely chiral algebras:

a) The sum of the algebraic matrix forms of two oppositely chiral algebras has zeros everywhere except on the top row, the left-most column and the leading diagonal; this is the 'oppositeness' of the minus signs referred to above.

b) Oppositely chiral algebras have algebraic matrix forms which commute.

Into 8-dimensions:

The reader is reminded, see (21.1), that:

a) The case 2 and the case 20 algebras share the same commutation relations.

b) The case 42 and case 60 algebras share the same commutation relations.

c) The commutation relations of the case 2 and case 20 algebras are opposite to the commutation relations of the case 42 and case 60 algebras.

Perhaps the chirality of two 8-dimensional algebraic matrix forms of opposite chirality also have an 'oppositeness' of signs as we found in the 4-dimensional algebras, (21.4). If we add the algebraic matrix forms of the case 2 and the case 60 algebras using the same variables, we get:

$$2\begin{bmatrix} a & b & c & d & e & f & g & h \\ -b & a & 0 & 0 & 0 & 0 & 0 & 0 \\ -c & 0 & a & 0 & 0 & 0 & 0 & 0 \\ -d & 0 & 0 & a & 0 & 0 & 0 & 0 \\ e & 0 & 0 & 0 & a & 0 & 0 & 0 \\ -f & 0 & 0 & 0 & 0 & a & 0 & 0 \\ -g & 0 & 0 & 0 & 0 & 0 & a & 0 \\ -h & 0 & 0 & 0 & 0 & 0 & 0 & a \end{bmatrix} \qquad (21.6)$$

This matrix is of the form of the sum of the two chiral opposite quaternions, (21.4), and so we might take it that the case 2 and the case 60 algebras are chiral opposite algebras. We get the same matrix, (21.6), if we add the case 20 and the case 42 algebraic matrix forms. Of course, what we are seeing here, (21.6), is that the distribution of minus signs in the 'body[104]' of the algebraic matrix forms of the two algebras are exactly opposite.

However, the case 2 and the case 60 algebras do not commute even if they have the same variables, and the case 20 and case 42 algebras do not commute even if they have the same variables. The commutator of the case 2 algebraic matrix form and the case 60 algebraic matrix form with the same variables is:

$$[C_2 C_{60} - C_{60} C_2] = 4 \begin{bmatrix} 0 & ef & eg & eh & -(bf+cg+dh) & be & ce & de \\ ef & 0 & 0 & 0 & 0 & 0 & 0 & 0 \\ eg & 0 & 0 & 0 & 0 & 0 & 0 & 0 \\ eh & 0 & 0 & 0 & 0 & 0 & 0 & 0 \\ bf+cg+dh & 0 & 0 & 0 & 0 & 0 & 0 & 0 \\ eb & 0 & 0 & 0 & 0 & 0 & 0 & 0 \\ ec & 0 & 0 & 0 & 0 & 0 & 0 & 0 \\ ed & 0 & 0 & 0 & 0 & 0 & 0 & 0 \end{bmatrix} \qquad (21.7)$$

The commutator of the case 20 algebraic matrix form and the case 42 algebraic matrix form sharing the same variables is the negative of this, (21.7):

$$[C_2 C_{60} - C_{60} C_2] = -[C_{20} C_{42} - C_{42} C_{20}] \qquad (21.8)$$

Other pairs of algebras in the other groups of four algebras sharing the same variables have sums of the same form as (21.6) and commutators of the same form as (21.7) other than a few minus signs. We have:

---

[104] We use the word 'body' to mean every element except the top row, the left-most column and the leading diagonal of the matrix.

$$[C_4 C_{58} - C_{58} C_4] = 4 \begin{bmatrix} 0 & ef & -eg & -eh & -bf+cg+dh & be & -ce & -de \\ ef & 0 & 0 & 0 & 0 & 0 & 0 & 0 \\ -eg & 0 & 0 & 0 & 0 & 0 & 0 & 0 \\ -eh & 0 & 0 & 0 & 0 & 0 & 0 & 0 \\ bf-cg-dh & 0 & 0 & 0 & 0 & 0 & 0 & 0 \\ eb & 0 & 0 & 0 & 0 & 0 & 0 & 0 \\ -ec & 0 & 0 & 0 & 0 & 0 & 0 & 0 \\ -ed & 0 & 0 & 0 & 0 & 0 & 0 & 0 \end{bmatrix} \quad (21.9)$$

Some pairs of algebras do commute:

However, the case 2 algebraic matrix form commutes with the case 42 algebraic matrix form (with different variables). Also, the case 20 algebraic matrix form commutes with the case 60 algebraic matrix form.

The above (21.7) & (21.9) shows that the lack of commutativity is not within the values of the variables. We have calculated these commutators, (21.7) & (21.9), deliberately with the same variables to show that the non-commutativity is a feature of the algebraic matrix forms and cannot be dispensed with for any choice of values of the variables. With different variables, the case 2 and the case 60 algebras are as completely non-commutative as is possible for two matrices.

If we add the case 2 and the case 42 algebraic matrix forms, we get:

$$\begin{bmatrix} a & b & c & d & e & f & g & h \\ -b & a & 0 & 0 & f & -e & 0 & 0 \\ -c & 0 & a & 0 & g & 0 & -e & 0 \\ -d & 0 & 0 & a & h & 0 & 0 & -e \\ e & -f & -g & -h & a & -b & -c & -d \\ -f & -e & 0 & 0 & b & a & 0 & 0 \\ -g & 0 & -e & 0 & c & 0 & a & 0 \\ -h & 0 & 0 & -e & d & 0 & 0 & a \end{bmatrix} \quad (21.10)$$

We get the same distribution of minus signs if we add the algebraic matrix forms of the case 20 and the case 60 algebras[105].

What we have discovered in the set of four algebras with case numbers $\{2, 20, 42, 60\}$ applies to the other sets of four algebras.

---

[105] There are some sign differences in the non-zero elements.

A complicated chirality:

We can separate the sixteen $Cl_{0,3} = 1 + 6\sqrt[2]{-1} + \sqrt[2]{+1}$ algebras into four sets of four algebras identified by case numbers:

$$\begin{Bmatrix} \{2,20\} \\ \{42,60\} \end{Bmatrix} \quad \begin{Bmatrix} \{4,18\} \\ \{44,58\} \end{Bmatrix} \quad \begin{Bmatrix} \{10,28\} \\ \{34,52\} \end{Bmatrix} \quad \begin{Bmatrix} \{12,26\} \\ \{36,50\} \end{Bmatrix} \tag{21.11}$$

The commutation relations of each bracketed pair of cases are identical. The commutation relations of the second pair in any pair of pairs of cases are opposite to the first pair.

The chirality relations of each set of four algebras is similar to the chirality relations of the four cases $\{2,20,42,60\}$ which are:

a) The chirality of case 42 is opposite to the chirality of case 2 in that these two algebraic matrix forms commute.
b) The chirality of case 60 is opposite to the chirality of case 20 in that these two algebraic matrix forms commute.
c) The chirality of case 42 is opposite to the chirality of case 20 in that these two algebraic matrix forms sum to a matrix of the form (21.6).
d) The chirality of case 60 is opposite to the chirality of case 2 in that these two algebraic matrix forms sum to a matrix of the form (21.6).

Remarkable!

A quandary:

We have a quandary. Do we define opposite chirality based on the commutation of two algebraic matrix forms or do we define opposite chirality of the basis of the distribution of minus signs within the 'body' of the matrix or do we define opposite chirality based on the direction of the commutation relations? In the 4-dimensional algebras, all three definitions give the same results. In 8-dimensions, the results are different. Both the case 42 and the case 60 algebras have opposite commutation relations to the case 2 algebra, but which algebra, case 42 or case 60, do we say is of opposite chirality to the case 2 algebra? Are they both opposite? We have a marriage of four rather than a marriage of two.

The commutation relations of the forty-eight $Cl_{2,1}$ algebras:

The commutation relations of the $Cl_{2,1} = 1 + 5\sqrt{+1} + 2\sqrt{-1}$ algebras															
Case	bc	bd	bf	bg	bh	cd	cf	cg	ch	df	dg	dh	fg	fh	gh
6	$-d$	$+c$	0	$-h$	$+g$	$+b$	$+h$	0	$+f$	$-g$	$-f$	0	$-d$	$+c$	$+b$
8	$-d$	$+c$	0	$-h$	$+g$	$+b$	$-h$	0	$-f$	$+g$	$+f$	0	$+d$	$-c$	$+b$
14	$-d$	$+c$	0	$+h$	$-g$	$+b$	$-h$	0	$-f$	$-g$	$-f$	0	$-d$	$-c$	$-b$
16	$-d$	$+c$	0	$+h$	$-g$	$+b$	$+h$	0	$+f$	$+g$	$+f$	0	$+d$	$+c$	$-b$
22	$-d$	$+c$	0	$-h$	$+g$	$+b$	$-h$	0	$-f$	$+g$	$+f$	0	$+d$	$-c$	$+b$

24	−d	+c	0	−h	+g	+b	+h	0	+f	−g	−f	0	−d	+c	+b
30	−d	+c	0	+h	−g	+b	+h	0	+f	+g	+f	0	+d	+c	−b
32	−d	+c	0	+h	−g	+b	−h	0	−f	−g	−f	0	−d	−c	−b
38	+d	−c	0	−h	+g	−b	+h	0	+f	+g	+f	0	+d	+c	+b
40	+d	−c	0	−h	+g	−b	−h	0	−f	−g	−f	0	−d	−c	+b
46	+d	−c	0	+h	−g	−b	−h	0	−f	+g	+f	0	+d	−c	−b
48	+d	−c	0	+h	−g	−b	+h	0	+f	−g	−f	0	−d	+c	−b
54	+d	−c	0	−h	+g	−b	−h	0	−f	−g	−f	0	−d	−c	+b
56	+d	−c	0	−h	+g	−b	+h	0	+f	+g	+f	0	+d	+c	+b
62	+d	−c	0	+h	−g	−b	+h	0	+f	−g	−f	0	−d	+c	−b
64	+d	−c	0	+h	−g	−b	−h	0	−f	+g	+f	0	+d	−c	−b
66	−d	−c	0	−h	−g	−b	+h	0	−f	+g	+f	0	−d	−c	−b
68	−d	−c	0	−h	−g	−b	−h	0	+f	−g	−f	0	+d	+c	−b
70	−d	−c	0	−h	−g	+b	+h	0	+f	+g	−f	0	−d	−c	+b
72	−d	−c	0	−h	−g	+b	−h	0	−f	−g	+f	0	+d	+c	+b
74	−d	−c	0	+h	+g	−b	−h	0	+f	+g	+f	0	−d	+c	+b
76	−d	−c	0	+h	+g	−b	+h	0	−f	−g	−f	0	+d	−c	+b
78	−d	−c	0	+h	+g	+b	−h	0	−f	+g	−f	0	−d	+c	−b
80	−d	−c	0	+h	+g	+b	+h	0	+f	−g	+f	0	+d	−c	−b
82	−d	−c	0	−h	−g	−b	−h	0	+f	−g	−f	0	+d	+c	−b
84	−d	−c	0	−h	−g	−b	+h	0	−f	+g	+f	0	−d	−c	−b
86	−d	−c	0	−h	−g	+b	−h	0	−f	−g	+f	0	+d	+c	+b
88	−d	−c	0	−h	−g	+b	+h	0	+f	+g	−f	0	−d	−c	+b
90	−d	−c	0	+h	+g	−b	+h	0	−f	−g	−f	0	+d	−c	+b
92	−d	−c	0	+h	+g	−b	−h	0	+f	+g	+f	0	−d	+c	+b
94	−d	−c	0	+h	+g	+b	+h	0	+f	−g	+f	0	+d	−c	−b
96	−d	−c	0	+h	+g	+b	−h	0	−f	+g	−f	0	−d	+c	−b
98	+d	+c	0	−h	−g	+b	+h	0	−f	−g	−f	0	+d	−c	−b
100	+d	+c	0	−h	−g	+b	−h	0	+f	+g	+f	0	−d	+c	−b
102	+d	+c	0	−h	−g	−b	+h	0	+f	−g	+f	0	+d	−c	+b
104	+d	+c	0	−h	−g	−b	−h	0	−f	+g	−f	0	−d	+c	+b
106	+d	+c	0	+h	+g	+b	−h	0	+f	−g	−f	0	+d	+c	+b
108	+d	+c	0	+h	+g	+b	+h	0	−f	+g	+f	0	−d	−c	+b
110	+d	+c	0	+h	+g	−b	−h	0	−f	−g	+f	0	+d	+c	−b
112	+d	+c	0	+h	+g	−b	+h	0	+f	+g	−f	0	−d	−c	−b
114	+d	+c	0	−h	−g	+b	−h	0	+f	+g	+f	0	−d	+c	−b
116	+d	+c	0	−h	−g	+b	+h	0	−f	−g	−f	0	+d	−c	−b
118	+d	+c	0	−h	−g	−b	−h	0	−f	+g	−f	0	−d	+c	+b
120	+d	+c	0	−h	−g	−b	+h	0	+f	−g	+f	0	+d	−c	+b
122	+d	+c	0	+h	+g	+b	+h	0	−f	+g	+f	0	−d	−c	+b
124	+d	+c	0	+h	+g	+b	−h	0	+f	−g	−f	0	+d	+c	+b
126	+d	+c	0	+h	+g	−b	+h	0	+f	+g	−f	0	−d	−c	−b

128	+d	+c	0	+h	+g	−b	−h	0	−f	−g	+f	0	+d	+c	−b

Examination of the commutation relations of the forty-eight $Cl_{2,1} = 1 + 5\sqrt{+1} + 2\sqrt{-1}$ algebras will show a pattern similar to the pattern we found in the sixteen $Cl_{0,3} = 1 + 6\sqrt[2]{-1} + \sqrt[2]{+1}$ algebras, (21.11). The algebras occur in pairs which have the same commutation relations. Corresponding to each pair, there is an oppositely chiral pair with exactly the opposite commutation relations. Using the case numbers to designate the algebras, those pairings are:

$$\left\{ \begin{matrix} \{6,24\} \\ \{46,64\} \end{matrix} \right\}, \left\{ \begin{matrix} \{8,22\} \\ \{48,62\} \end{matrix} \right\}, \left\{ \begin{matrix} \{14,32\} \\ \{38,56\} \end{matrix} \right\}, \left\{ \begin{matrix} \{16,30\} \\ \{40,54\} \end{matrix} \right\}, \left\{ \begin{matrix} \{66,84\} \\ \{106,124\} \end{matrix} \right\}, \left\{ \begin{matrix} \{68,82\} \\ \{108,122\} \end{matrix} \right\}$$

$$\left\{ \begin{matrix} \{70,88\} \\ \{100,128\} \end{matrix} \right\}, \left\{ \begin{matrix} \{72,86\} \\ \{112,126\} \end{matrix} \right\}, \left\{ \begin{matrix} \{74,92\} \\ \{98,116\} \end{matrix} \right\}, \left\{ \begin{matrix} \{76,90\} \\ \{100,114\} \end{matrix} \right\}, \left\{ \begin{matrix} \{78,96\} \\ \{102,120\} \end{matrix} \right\}, \left\{ \begin{matrix} \{80,94\} \\ \{104,118\} \end{matrix} \right\}$$

$$(21.12)$$

The case 8 algebraic matrix form commutes with the case 48 algebraic matrix form. The case 22 algebraic matrix form commutes with the case 62 algebraic matrix form. The sum of the algebraic matrix forms of the case 8 algebra and the case 62 algebra is of the form (21.6) as is the sum of the algebraic matrix forms of the case 22 algebra and the case 48 algebra. There are similar relations for the other sets of four algebras in (21.12).

The commutation relations of the sixty-four $Cl_{3,0}$ algebras:

The commutation relations of the $Cl_{3,0} \cong Cl_{1,2} = 1 + 3\sqrt{+1} + 4\sqrt{-1}\,_{Non-Com}$ algebras															
Case	bc	bd	bf	bg	bh	cd	cf	cg	ch	df	dg	dh	fg	fh	gh
1	−d	+c	0	−h	+g	−b	+h	0	−f	−g	+f	0	+d	−c	+b
3	−d	+c	0	−h	+g	−b	−h	0	+f	+g	−f	0	−d	+c	+b
5	−d	+c	0	−h	+g	+b	+h	0	+f	−g	−f	0	+d	−c	−b
7	−d	+c	0	−h	+g	+b	−h	0	−f	+g	+f	0	−d	+c	−b
9	−d	+c	0	+h	−g	−b	−h	0	+f	−g	+f	0	+d	+c	−b
11	−d	+c	0	+h	−g	−b	+h	0	−f	+g	−f	0	−d	−c	−b
13	−d	+c	0	+h	−g	+b	−h	0	−f	−g	−f	0	+d	+c	+b
15	−d	+c	0	+h	−g	+b	+h	0	+f	+g	+f	0	−d	−c	+b
17	−d	+c	0	−h	+g	−b	−h	0	+f	+g	−f	0	−d	+c	+b
19	−d	+c	0	−h	+g	−b	+h	0	−f	−g	+f	0	+d	−c	+b
21	−d	+c	0	−h	+g	+b	−h	0	−f	+g	+f	0	−d	+c	−b
23	−d	+c	0	−h	+g	+b	+h	0	+f	−g	−f	0	+d	−c	−b
25	−d	+c	0	+h	−g	−b	+h	0	−f	+g	−f	0	−d	−c	−b
27	−d	+c	0	+h	−g	−b	−h	0	+f	−g	+f	0	+d	+c	−b
29	−d	+c	0	+h	−g	+b	+h	0	+f	+g	+f	0	−d	−c	+b
31	−d	+c	0	+h	−g	+b	−h	0	−f	−g	−f	0	+d	+c	+b
33	+d	−c	0	−h	+g	+b	+h	0	−f	+g	−f	0	−d	−c	+b
35	+d	−c	0	−h	+g	+b	−h	0	+f	−g	+f	0	+d	+c	+b
37	+d	−c	0	−h	+g	−b	+h	0	+f	+g	+f	0	−d	−c	−b

39	+d	−c	0	−h	+g	−b	−h	0	−f	−g	−f	0	+d	+c	−b
41	+d	−c	0	+h	−g	+b	−h	0	+f	+g	−f	0	−d	+c	−b
43	+d	−c	0	+h	−g	+b	+h	0	−f	−g	+f	0	+d	−c	−b
45	+d	−c	0	+h	−g	−b	−h	0	−f	+g	+f	0	−d	+c	+b
47	+d	−c	0	+h	−g	−b	+h	0	+f	−g	−f	0	+d	−c	+b
49	+d	−c	0	−h	+g	+b	−h	0	+f	−g	+f	0	+d	+c	+b
51	+d	−c	0	−h	+g	+b	+h	0	−f	+g	−f	0	−d	−c	+b
53	+d	−c	0	−h	+g	−b	−h	0	−f	−g	−f	0	+d	+c	−b
55	+d	−c	0	−h	+g	−b	+h	0	+f	+g	+f	0	−d	−c	−b
57	+d	−c	0	+h	−g	+b	+h	0	−f	−g	+f	0	+d	−c	−b
59	+d	−c	0	+h	−g	+b	−h	0	+f	+g	−f	0	−d	+c	−b
61	+d	−c	0	+h	−g	−b	+h	0	+f	−g	−f	0	+d	−c	+b
63	+d	−c	0	+h	−g	−b	−h	0	−f	+g	+f	0	−d	+c	+b
65	−d	−c	0	−h	−g	−b	+h	0	−f	+g	+f	0	+d	+c	+b
67	−d	−c	0	−h	−g	−b	−h	0	+f	−g	−f	0	−d	−c	+b
69	−d	−c	0	−h	−g	+b	+h	0	+f	+g	−f	0	+d	+c	−b
71	−d	−c	0	−h	−g	+b	−h	0	−f	−g	+f	0	−d	−c	−b
73	−d	−c	0	+h	+g	−b	−h	0	+f	+g	+f	0	+d	−c	−b
75	−d	−c	0	+h	+g	−b	+h	0	−f	−g	−f	0	−d	+c	−b
77	−d	−c	0	+h	+g	+b	−h	0	−f	+g	−f	0	+d	−c	+b
79	−d	−c	0	+h	+g	+b	+h	0	+f	−g	+f	0	−d	+c	+b
81	−d	−c	0	−h	−g	−b	−h	0	+f	−g	−f	0	−d	−c	+b
83	−d	−c	0	−h	−g	−b	+h	0	−f	+g	+f	0	+d	+c	+b
85	−d	−c	0	−h	−g	+b	−h	0	−f	−g	+f	0	−d	−c	−b
87	−d	−c	0	−h	−g	+b	+h	0	+f	+g	−f	0	+d	+c	−b
89	−d	−c	0	+h	+g	−b	+h	0	−f	−g	−f	0	−d	+c	−b
91	−d	−c	0	+h	+g	−b	−h	0	+f	+g	+f	0	+d	−c	−b
93	−d	−c	0	+h	+g	+b	+h	0	+f	−g	+f	0	−d	+c	+b
95	−d	−c	0	+h	+g	+b	−h	0	−f	+g	−f	0	+d	−c	+b
97	+d	+c	0	−h	−g	+b	+h	0	−f	−g	−f	0	−d	+c	+b
99	+d	+c	0	−h	−g	+b	−h	0	+f	+g	+f	0	+d	−c	+b
101	+d	+c	0	−h	−g	−b	+h	0	+f	−g	+f	0	−d	+c	−b
103	+d	+c	0	−h	−g	−b	−h	0	−f	+g	−f	0	+d	−c	−b
105	+d	+c	0	+h	+g	+b	−h	0	+f	−g	−f	0	−d	−c	−b
107	+d	+c	0	+h	+g	+b	+h	0	−f	+g	+f	0	+d	+c	−b
109	+d	+c	0	+h	+g	−b	−h	0	−f	−g	+f	0	−d	−c	+b
111	+d	+c	0	+h	+g	−b	+h	0	+f	+g	−f	0	+d	+c	+b
113	+d	+c	0	−h	−g	+b	−h	0	+f	+g	+f	0	+d	−c	+b
115	+d	+c	0	−h	−g	+b	+h	0	−f	−g	−f	0	−d	+c	+b
117	+d	+c	0	−h	−g	−b	−h	0	−f	+g	−f	0	+d	−c	−b
119	+d	+c	0	−h	−g	−b	+h	0	+f	−g	+f	0	−d	+c	−b
121	+d	+c	0	+h	+g	+b	+h	0	−f	+g	+f	0	+d	+c	−b
123	+d	+c	0	+h	+g	+b	−h	0	+f	−g	−f	0	−d	−c	−b

125	+d	+c	0	+h	+g	−b	+h	0	+f	+g	−f	0	+d	+c	+b
127	+d	+c	0	+h	+g	−b	−h	0	−f	−g	+f	0	−d	−c	+b

Examination of the commutation relations of the sixty-four $Cl_{3,0} \cong Cl_{1,2} = 1 + 3\sqrt{+1} + 4\sqrt{-1}_{Non-Com}$ algebras shows a pattern similar to the pattern we found in the sixteen $Cl_{0,3} = 1 + 6\sqrt[2]{-1} + \sqrt[2]{+1}$ algebras, (21.11). The algebras occur in pairs which have the same commutation relations. Corresponding to each pair, there is an oppositely chiral pair with exactly the opposite commutation relations. Using the case numbers to designate the algebras, those pairings are:

$$\left\{\begin{matrix}\{1,19\}\\\{41,59\}\end{matrix}\right\}, \left\{\begin{matrix}\{3,17\}\\\{43,57\}\end{matrix}\right\}, \left\{\begin{matrix}\{5,23\}\\\{45,63\}\end{matrix}\right\}, \left\{\begin{matrix}\{7,21\}\\\{47,61\}\end{matrix}\right\}, \left\{\begin{matrix}\{9,27\}\\\{33,51\}\end{matrix}\right\}, \left\{\begin{matrix}\{11,25\}\\\{35,49\}\end{matrix}\right\}$$

$$\left\{\begin{matrix}\{13,31\}\\\{37,55\}\end{matrix}\right\}, \left\{\begin{matrix}\{15,29\}\\\{39,53\}\end{matrix}\right\}, \left\{\begin{matrix}\{65,83\}\\\{105,123\}\end{matrix}\right\}, \left\{\begin{matrix}\{67,81\}\\\{107,121\}\end{matrix}\right\}, \left\{\begin{matrix}\{69,87\}\\\{109,127\}\end{matrix}\right\} \qquad (21.13)$$

$$\left\{\begin{matrix}\{71,85\}\\\{111,125\}\end{matrix}\right\}, \left\{\begin{matrix}\{73,91\}\\\{97,115\}\end{matrix}\right\}, \left\{\begin{matrix}\{75,89\}\\\{99,113\}\end{matrix}\right\}, \left\{\begin{matrix}\{77,95\}\\\{101,119\}\end{matrix}\right\}, \left\{\begin{matrix}\{79,93\}\\\{103,117\}\end{matrix}\right\}$$

The case 1 algebraic matrix form commutes with the case 41 algebraic matrix form. The case 19 algebraic matrix form commutes with the case 59 algebraic matrix form. The sum of the algebraic matrix forms of the case 1 algebra and the case 59 algebra is of the form (21.6) as is the sum of the algebraic matrix forms of the case 19 algebra and the case 41 algebra. There are similar relations for the other sets of four algebras in (21.13).

Chapter 22

Married Algebras

We will consider the four 'married' algebras:

$$\left\{ \begin{array}{l} \{2,20\} \\ \{42,60\} \end{array} \right\} \tag{22.1}$$

We have:

$Cl_{0,3} = 1 + 1\sqrt{+1} + 6\sqrt{-1}$							
Case Number	$P_{2,1}$	$P_{2,4}$	$P_{2,6}$	$P_{2,8}$	$P_{3,1}$	$P_{3,7}$	$P_{5,1}$
2	-1	-1	-1	-1	-1	-1	
20	-1	-1		-1	-1		
42	-1		-1		-1	-1	
60	-1				-1		

We have Case 2:

$$\begin{bmatrix} a & b & c & d & e & f & g & h \\ -b & a & d & -c & f & -e & h & -g \\ -c & -d & a & b & g & -h & -e & f \\ -d & c & -b & a & h & g & -f & -e \\ e & -f & -g & -h & a & -b & -c & -d \\ -f & -e & h & -g & b & a & d & -c \\ -g & -h & -e & f & c & -d & a & b \\ -h & g & -f & -e & d & c & -b & a \end{bmatrix} \equiv \begin{bmatrix} \left[\mathbb{H}_{R\chi} \right] & \sim \\ \sim & \left[\mathbb{H}_{L\chi} \right] \end{bmatrix} \tag{22.2}$$

Case 2: $P_{2,1} = P_{2,4} = P_{2,6} = P_{2,8} = P_{3,1} = P_{3,7} = -1, \qquad P_{5,1} = +1$

Looking at this, (22.2), we see that the top right-hand 4×4 corner is not a sub-algebra. The signs of the e variable would have to be the same sign for this to be of the form of a 4-dimensional algebra. Neither is the bottom left-hand 4×4 corner a sub-algebra.

The 4-dimensional sub-algebras of every $C_2 \times C_2 \times C_2$ algebra are:

$$\begin{array}{c} \{a,b,c,d\}, \quad \{a,b,e,f\}, \quad \{a,b,g,h\}, \quad \{a,c,e,g\} \\ \{a,c,f,h\}, \quad \{a,d,e,h\}, \quad \{a,d,f,g\} \end{array} \tag{22.3}$$

It will never be the case that we have sub-algebras in the off diagonal 4×4 corners. However, as we see in the case 20 algebra, these off diagonal 4×4 corners can take the form of 4-dimensional algebras.

We have case 20:

$$
\begin{bmatrix}
a & b & c & d & e & f & g & h \\
-b & a & d & -c & -f & e & h & -g \\
-c & -d & a & b & -g & -h & e & f \\
-d & c & -b & a & -h & g & -f & e \\
e & f & g & h & a & b & c & d \\
-f & e & h & -g & -b & a & d & -c \\
-g & -h & e & f & -c & -d & a & b \\
-h & g & -f & e & -d & c & -b & a
\end{bmatrix}
\equiv
\begin{bmatrix}
\begin{bmatrix} \mathbb{H}_{R\chi} \end{bmatrix} & \begin{bmatrix} \mathbb{H}_{R\chi} \end{bmatrix} \\
\begin{bmatrix} \mathbb{H}_{R\chi} \end{bmatrix} & \begin{bmatrix} \mathbb{H}_{R\chi} \end{bmatrix}
\end{bmatrix}
$$

(22.4)

Case 20 : $P_{2,1} = P_{2,4} = P_{2,8} = P_{3,1} = -1, \qquad P_{2,6} = P_{3,7} = P_{5,1} = +1$

We have case 42:

$$
\begin{bmatrix}
a & b & c & d & e & f & g & h \\
-b & a & -d & c & f & -e & -h & g \\
-c & d & a & -b & g & h & -e & -f \\
-d & -c & b & a & h & -g & f & -e \\
e & -f & -g & -h & a & -b & -c & -d \\
-f & -e & -h & g & b & a & -d & c \\
-g & h & -e & -f & c & d & a & -b \\
-h & -g & f & -e & d & -c & b & a
\end{bmatrix}
\equiv
\begin{bmatrix}
\begin{bmatrix} \mathbb{H}_{L\chi} \end{bmatrix} & \sim \\
\sim & \begin{bmatrix} \mathbb{H}_{R\chi} \end{bmatrix}
\end{bmatrix}
$$

(22.5)

Case 42 : $P_{2,1} = P_{2,6} = P_{3,1} = P_{3,7} = -1, \qquad P_{2,4} = P_{2,8} = P_{5,1} = +1$

We have case 60:

$$
\begin{bmatrix}
a & b & c & d & e & f & g & h \\
-b & a & -d & c & -f & e & -h & g \\
-c & d & a & -b & -g & h & e & -f \\
-d & -c & b & a & -h & -g & f & e \\
e & f & g & h & a & b & c & d \\
-f & e & -h & g & -b & a & -d & c \\
-g & h & e & -f & -c & d & a & -b \\
-h & -g & f & e & -d & -c & b & a
\end{bmatrix}
\equiv
\begin{bmatrix}
\begin{bmatrix} \mathbb{H}_{L\chi} \end{bmatrix} & \begin{bmatrix} \mathbb{H}_{L\chi} \end{bmatrix} \\
\begin{bmatrix} \mathbb{H}_{L\chi} \end{bmatrix} & \begin{bmatrix} \mathbb{H}_{L\chi} \end{bmatrix}
\end{bmatrix}
$$

(22.6)

Case 60 : $P_{2,1} = P_{3,1} = -1, \qquad P_{2,4} = P_{2,6} = P_{2,8} = P_{3,7} = P_{5,1} = +1$

Matrices have the property of block multiplication. Writing two 8×8 matrices as four 4×4 blocks and multiplying the two block matrices together using matrix multiplication as if we had two 2×2 matrices will produce the same product as multiplying the two 8×8 matrices together.

Looking at (22.4) & (22.6), we can see why these matrices are chiral opposites and why they commute.

The cases 20 and 60 are particularly clear. If we take the set of algebras:

$$\begin{Bmatrix} \{4,18\} \\ \{44,58\} \end{Bmatrix} \tag{22.7}$$

We have:

$$Case\ 4 = \begin{bmatrix} \mathbb{H}_{R\chi} & \sim \\ \sim & \mathbb{H}_{L\chi} \end{bmatrix} \qquad Case\ 18 = \begin{bmatrix} \mathbb{H}_{R\chi} & \sim \\ \sim & \mathbb{H}_{R\chi} \end{bmatrix}$$

$$Case\ 44 = \begin{bmatrix} \mathbb{H}_{L\chi} & \sim \\ \sim & \mathbb{H}_{R\chi} \end{bmatrix} \qquad Case\ 58 = \begin{bmatrix} \mathbb{H}_{L\chi} & \sim \\ \sim & \mathbb{H}_{L\chi} \end{bmatrix} \tag{22.8}$$

We see that the copies of $\mathbb{H}_{R\chi}$ in the top left-hand corners of case 4 and case 18 will ensure that these two algebras do not commute. We also see that the copies of $\mathbb{H}$ in the case 4 and case 44 and in the case 18 and case 58 will allow that these two pairs of algebras can commute if the 'squiggle corners' do not prevent commutation.

The 'squiggle corners':
Consider the upper right-hand 'squiggle corner' in case 2 algebra, (22.2). We have:

$$Case\ 2: \qquad \sim_{Top} = \begin{bmatrix} e & f & g & h \\ f & -e & h & -g \\ g & -h & -e & f \\ h & g & -f & -e \end{bmatrix} \tag{22.9}$$

We have:

$$\begin{bmatrix} 1 & 0 & 0 & 0 \\ 0 & -1 & 0 & 0 \\ 0 & 0 & -1 & 0 \\ 0 & 0 & 0 & -1 \end{bmatrix} \begin{bmatrix} e & f & g & h \\ f & -e & h & -g \\ g & -h & -e & f \\ h & g & -f & -e \end{bmatrix} = \begin{bmatrix} e & f & g & h \\ -f & e & -h & g \\ -g & h & e & -f \\ -h & -g & f & e \end{bmatrix} = \mathbb{H}_{L\chi} \tag{22.10}$$

And:

$$\begin{bmatrix} e & f & g & h \\ f & -e & h & -g \\ g & -h & -e & f \\ h & g & -f & -e \end{bmatrix} \begin{bmatrix} 1 & 0 & 0 & 0 \\ 0 & -1 & 0 & 0 \\ 0 & 0 & -1 & 0 \\ 0 & 0 & 0 & -1 \end{bmatrix} = \begin{bmatrix} e & -f & -g & -h \\ f & e & -h & g \\ g & h & e & -f \\ h & -g & f & e \end{bmatrix} = \mathbb{H}^{*}_{R\chi} \tag{22.11}$$

We have used the * sign to indicate conjugation (reversal of the signs of the imaginary variables in the quaternion algebras).

Consider the lower left-hand 'squiggle corner' in case 2 algebra, (22.2). We have:

$$Case\ 2: \qquad \sim_{Bottom} = \begin{bmatrix} e & -f & -g & -h \\ -f & -e & h & -g \\ -g & -h & -e & f \\ -h & g & -f & -e \end{bmatrix} \qquad (22.12)$$

We have:

$$\begin{bmatrix} 1 & 0 & 0 & 0 \\ 0 & -1 & 0 & 0 \\ 0 & 0 & -1 & 0 \\ 0 & 0 & 0 & -1 \end{bmatrix} \begin{bmatrix} e & -f & -g & -h \\ -f & -e & h & -g \\ -g & -h & -e & f \\ -h & g & -f & -e \end{bmatrix} = \begin{bmatrix} e & -f & -g & -h \\ f & e & -h & g \\ g & h & e & -f \\ h & -g & f & e \end{bmatrix} = \mathbb{H}^*_{R\chi} \qquad (22.13)$$

And:

$$\begin{bmatrix} e & -f & -g & -h \\ -f & -e & h & -g \\ -g & -h & -e & f \\ -h & g & -f & -e \end{bmatrix} \begin{bmatrix} 1 & 0 & 0 & 0 \\ 0 & -1 & 0 & 0 \\ 0 & 0 & -1 & 0 \\ 0 & 0 & 0 & -1 \end{bmatrix} = \begin{bmatrix} e & f & g & h \\ -f & e & -h & g \\ -g & h & e & -f \\ -h & -g & f & e \end{bmatrix} = \mathbb{H}_{L\chi} \qquad (22.14)$$

This allows us to write the case 2 algebra as:

$$Case\ 2 = \begin{bmatrix} \mathbb{H}_{R\chi} & M\mathbb{H}_{L\chi} \\ M\mathbb{H}^*_{R\chi} & \mathbb{H}_{L\chi} \end{bmatrix} = \begin{bmatrix} \mathbb{H}_{R\chi} & \mathbb{H}_{R\chi}M \\ \mathbb{H}^*_{L\chi}M & \mathbb{H}_{L\chi} \end{bmatrix} \quad where \quad M = \begin{bmatrix} 1 & 0 & 0 & 0 \\ 0 & -1 & 0 & 0 \\ 0 & 0 & -1 & 0 \\ 0 & 0 & 0 & -1 \end{bmatrix} \qquad (22.15)$$

Similarly, the case 42 algebra can be written as:

$$Case\ 42 = \begin{bmatrix} \mathbb{H}_{L\chi} & M\mathbb{H}_{R\chi} \\ M\mathbb{H}^*_{L\chi} & \mathbb{H}_{R\chi} \end{bmatrix} = \begin{bmatrix} \mathbb{H}_{L\chi} & \mathbb{H}_{L\chi}M \\ \mathbb{H}^*_{R\chi}M & \mathbb{H}_{R\chi} \end{bmatrix} \quad where \quad M = \begin{bmatrix} 1 & 0 & 0 & 0 \\ 0 & -1 & 0 & 0 \\ 0 & 0 & -1 & 0 \\ 0 & 0 & 0 & -1 \end{bmatrix} \qquad (22.16)$$

The product of the case 2 algebra and the case 42 algebra is:

$$\begin{bmatrix} \mathbb{H}_{R\chi} & M\mathbb{H}_{L\chi} \\ M\mathbb{H}^*_{R\chi} & \mathbb{H}_{L\chi} \end{bmatrix}_2 \begin{bmatrix} \mathbb{H}_{L\chi} & \mathbb{H}_{L\chi}M \\ \mathbb{H}^*_{R\chi}M & \mathbb{H}_{R\chi} \end{bmatrix}_{42} = \begin{bmatrix} \mathbb{H}_{R\chi}\mathbb{H}_{L\chi} + M\mathbb{H}_{L\chi}\mathbb{H}^*_{R\chi}M & \mathbb{H}_{R\chi}\mathbb{H}_{L\chi}M + M\mathbb{H}_{L\chi}\mathbb{H}_{R\chi} \\ M\mathbb{H}^*_{R\chi}\mathbb{H}_{L\chi} + \mathbb{H}_{L\chi}\mathbb{H}^*_{R\chi}M & \mathbb{H}_{L\chi}\mathbb{H}_{R\chi} + M\mathbb{H}^*_{R\chi}\mathbb{H}_{L\chi}M \end{bmatrix}$$
$$(22.17)$$

We use the alternative matrices in (22.15) & (22.16) and reverse the order to get:

$$\begin{bmatrix} \mathbb{H}_{L\chi} & M\mathbb{H}_{R\chi} \\ M\mathbb{H}_{L\chi}^* & \mathbb{H}_{R\chi} \end{bmatrix}_{42} \begin{bmatrix} \mathbb{H}_{R\chi} & \mathbb{H}_{R\chi}M \\ \mathbb{H}_{L\chi}^*M & \mathbb{H}_{L\chi} \end{bmatrix}_2 = \begin{bmatrix} \mathbb{H}_{L\chi}\mathbb{H}_{R\chi} + M\mathbb{H}_{R\chi}\mathbb{H}_{L\chi}^*M & \mathbb{H}_{L\chi}\mathbb{H}_{R\chi}M + M\mathbb{H}_{R\chi}\mathbb{H}_{L\chi} \\ M\mathbb{H}_{L\chi}^*\mathbb{H}_{R\chi} + \mathbb{H}_{R\chi}\mathbb{H}_{L\chi}^*M & \mathbb{H}_{R\chi}\mathbb{H}_{L\chi} + M\mathbb{H}_{L\chi}^*\mathbb{H}_{R\chi}M \end{bmatrix}$$

$$(22.18)$$

Using:

$$\mathbb{H}_{L\chi}\mathbb{H}_{R\chi} = \mathbb{H}_{R\chi}\mathbb{H}_{L\chi} \qquad (22.19)$$

We see that these matrices, (22.17) & (22.18) are the same. The case 2 algebra commutes with the case 42 algebra.

Differentiation

Differentiation within division algebras, spinor algebras, is done with a differential operator. This is different from differentiation within a $\mathbb{R}^n$ space which is just n copies of the real numbers. A differential operator is just a convenient way of calculating; there is really no such thing as a differential operator within a division algebra. Each of the 128 algebras we will examine are each a separate distinct algebra in their own right. Each copy of each algebra has its own differential operator.

The differential operator is formed as the sum of the inverses of each variable within the particular algebra wherein the variables are replaced by a differential operator, ∂x_i. For example, the left-chiral quaternion differential operator is:

$$d^{\mathbb{H}_{Lx}} = \left\{ \begin{bmatrix} \partial t & 0 & 0 & 0 \\ 0 & \partial t & 0 & 0 \\ 0 & 0 & \partial t & 0 \\ 0 & 0 & 0 & \partial t \end{bmatrix} + \begin{bmatrix} 0 & -\partial x & 0 & 0 \\ \partial x & 0 & 0 & 0 \\ 0 & 0 & 0 & \partial x \\ 0 & 0 & -\partial x & 0 \end{bmatrix} + \begin{bmatrix} 0 & 0 & -\partial y & 0 \\ 0 & 0 & 0 & -\partial y \\ \partial y & 0 & 0 & 0 \\ 0 & \partial y & 0 & 0 \end{bmatrix} + \begin{bmatrix} 0 & 0 & 0 & -\partial z \\ 0 & 0 & \partial z & 0 \\ 0 & -\partial z & 0 & 0 \\ \partial z & 0 & 0 & 0 \end{bmatrix} \right\} \tag{23.1}$$

Beware, the sum of the inverses of each separate element is not the inverse of the algebraic matrix form.

The potential field:

In accordance with the quaternions, we seek E-fields which are of the form of space-time curls like the electric field. We are therefore tempted to reverse the sign of the spatial components and use potentials of the form:

$$\begin{bmatrix} \phi & -A_b & -A_c & -A_d & -A_e & -A_f & -A_g & -A_h \end{bmatrix} \tag{23.2}$$

However, the Clifford algebra tri-vector, which is the e variable in our algebraic matrix form, is special. It is a little like a 'second identity' in that it commutes with every other variable. We choose to treat the e variable in a way similar to how we treat the real variable. The e variable is not real of course. We point out that while the real variable within the 8-dimensional $C_2 \times C_2 \times C_2$ algebras can take only positive values, the e variable can take both positive and negative values.

We therefore use potentials of the form:

$$\begin{bmatrix} \phi_a & -A_b & -A_c & -A_d & \phi_e & -A_f & -A_g & -A_h \end{bmatrix} \tag{23.3}$$

We have changed the name of the A_e variable – no great change there; we have also changed the sign of the A_e variable- perhaps this is a great change. We opine that we are not 'rigging' the result here. The reader will form their own opinion regarding how decently we are proceeding, but, before contemning us, just look at the results.

The 8-dimensional differential operator:

The non-commutative 8-dimensional general differential operator is:

$$d = \begin{bmatrix}
\partial t & \partial b \dfrac{1}{P_{2,1}} & \partial c \dfrac{1}{P_{3,1}} & -\partial d \dfrac{P_{2,4}^2}{P_{2,1}P_{3,1}} & \partial e \dfrac{1}{P_{5,1}} & \partial f \dfrac{P_{2,6}^2}{P_{2,1}P_{5,1}} & \partial g \dfrac{P_{3,7}^2}{P_{3,1}P_{5,1}} & -\partial h \dfrac{P_{2,8}^2 P_{3,7}^2}{P_{2,1}P_{3,1}P_{5,1}} \\[2.5em]
\partial b & \partial t & -\partial d \dfrac{P_{2,4}}{P_{3,1}} & \partial c \dfrac{P_{2,4}}{P_{3,1}} & \partial f \dfrac{P_{2,6}}{P_{5,1}} & \partial e \dfrac{P_{2,6}}{P_{5,1}} & -\partial h \dfrac{P_{2,8}P_{3,7}^2}{P_{3,1}P_{5,1}} & \partial g \dfrac{P_{2,8}P_{3,7}^2}{P_{3,1}P_{5,1}} \\[2.5em]
\partial c & \partial d \dfrac{P_{2,4}}{P_{2,1}} & \partial t & -\partial b \dfrac{P_{2,4}}{P_{2,1}} & \partial g \dfrac{P_{3,7}}{P_{5,1}} & \partial h \dfrac{P_{2,6}P_{2,8}P_{3,7}}{P_{2,1}P_{5,1}} & \partial e \dfrac{P_{3,7}}{P_{5,1}} & -\partial f \dfrac{P_{2,6}P_{2,8}P_{3,7}}{P_{2,1}P_{5,1}} \\[2.5em]
\partial d & \partial c \dfrac{1}{P_{2,4}} & -\partial b \dfrac{1}{P_{2,4}} & \partial t & \partial h \dfrac{P_{2,8}P_{3,7}}{P_{2,4}P_{5,1}} & \partial g \dfrac{P_{2,6}P_{3,7}}{P_{2,4}P_{5,1}} & -\partial f \dfrac{P_{2,6}P_{3,7}}{P_{2,4}P_{5,1}} & \partial e \dfrac{P_{2,8}P_{3,7}}{P_{2,4}P_{5,1}} \\[2.5em]
\partial e & \partial f \dfrac{P_{2,6}}{P_{2,1}} & \partial g \dfrac{P_{3,7}}{P_{3,1}} & -\partial h \dfrac{P_{2,4}P_{2,8}P_{3,7}}{P_{2,1}P_{3,1}} & \partial t & \partial b \dfrac{P_{2,6}}{P_{2,1}} & \partial c \dfrac{P_{3,7}}{P_{3,1}} & -\partial d \dfrac{P_{2,4}P_{2,8}P_{3,7}}{P_{2,1}P_{3,1}} \\[2.5em]
\partial f & \partial e \dfrac{1}{P_{2,6}} & -\partial h \dfrac{P_{2,8}P_{3,7}}{P_{2,6}P_{3,1}} & \partial g \dfrac{P_{2,4}P_{3,7}}{P_{2,6}P_{3,1}} & \partial b \dfrac{1}{P_{2,6}} & \partial t & -\partial d \dfrac{P_{2,4}P_{3,7}}{P_{2,6}P_{3,1}} & \partial c \dfrac{P_{2,8}P_{3,7}}{P_{2,6}P_{3,1}} \\[2.5em]
\partial g & \partial h \dfrac{P_{2,8}}{P_{2,1}} & \partial e \dfrac{1}{P_{3,7}} & -\partial f \dfrac{P_{2,4}P_{2,6}}{P_{2,1}P_{3,7}} & \partial c \dfrac{1}{P_{3,7}} & \partial d \dfrac{P_{2,4}P_{2,6}}{P_{2,1}P_{3,7}} & \partial t & -\partial b \dfrac{P_{2,8}}{P_{2,1}} \\[2.5em]
\partial h & \partial g \dfrac{1}{P_{2,8}} & -\partial f \dfrac{P_{2,6}}{P_{2,8}P_{3,7}} & \partial e \dfrac{P_{2,4}}{P_{2,8}P_{3,7}} & \partial d \dfrac{P_{2,4}}{P_{2,8}P_{3,7}} & \partial c \dfrac{P_{2,6}}{P_{2,8}P_{3,7}} & -\partial b \dfrac{1}{P_{2,8}} & \partial t
\end{bmatrix}$$

$$\tag{23.4}$$

The case 2 algebra – an example:

In the case 2 $Cl_{0,3} = 1 + 6\sqrt[2]{-1} + \sqrt[2]{+1}$ algebra.

Case Number	$P_{2,1}$	$P_{2,4}$	$P_{2,6}$	$P_{2,8}$	$P_{3,1}$	$P_{3,7}$	$P_{5,1}$
2	−1	−1	−1	−1	−1	−1	

The potential of this particular algebra is:

$$\Phi_{Case\ 2} = \begin{bmatrix} \phi_a & -A_b & -A_c & -A_d & \phi_e & -A_f & -A_g & -A_h \\ A_b & \phi_a & -A_d & A_c & -A_f & -\phi_e & -A_h & A_g \\ A_c & A_d & \phi_a & -A_b & -A_g & A_h & -\phi_e & -A_f \\ A_d & -A_c & A_b & \phi_a & -A_h & -A_g & A_f & -\phi_e \\ \phi_e & A_f & A_g & A_h & \phi_a & A_b & A_c & A_d \\ A_f & -\phi_e & -A_h & A_g & -A_b & \phi_a & -A_d & A_c \\ A_g & A_h & -\phi_e & -A_f & -A_c & A_d & \phi_a & -A_b \\ A_h & -A_g & A_f & -\phi_e & -A_d & -A_c & A_b & \phi_a \end{bmatrix} \tag{23.5}$$

The differential operator of this particular algebra is:

$$d = \begin{bmatrix} \partial t & -\partial b & -\partial c & -\partial d & \partial e & -\partial f & -\partial g & -\partial h \\ \partial b & \partial t & -\partial d & \partial c & -\partial f & -\partial e & -\partial h & \partial g \\ \partial c & \partial d & \partial t & -\partial b & -\partial g & \partial h & -\partial e & -\partial f \\ \partial d & -\partial c & \partial b & \partial t & -\partial h & -\partial g & \partial f & -\partial e \\ \partial e & \partial f & \partial g & \partial h & \partial t & \partial b & \partial c & \partial d \\ \partial f & -\partial e & -\partial h & \partial g & -\partial b & \partial t & -\partial d & \partial c \\ \partial g & \partial h & -\partial e & -\partial f & -\partial c & \partial d & \partial t & -\partial b \\ \partial h & -\partial g & \partial f & -\partial e & -\partial d & -\partial c & \partial b & \partial t \end{bmatrix} \tag{23.6}$$

The E-fields of the 8-dimensional Algebras

There are only two types of non-commutative algebras which derive from the $C_2 \times C_2$ group; these are the quaternions (two algebras) and the A_3 algebras (six algebras). We formed our 4-dimensional space-time as the superimposition (sum) of all six A_3 algebras. We formed the symmetrical classical energy momentum tensor and the anti-symmetrical classical electromagnetic tensor as the superimposition (sum) of all six E-fields and all six B-fields of the A_3 algebras.

We formed the electron field as the superimposition of the two quaternion E-fields, and we formed the neutrino field as the superimposition of the two quaternion B-fields. We will treat the 8-dimensional algebras in a similar way. Since there are three distinct types of 8-dimensional $C_2 \times C_2$ algebras, we will get three '8-dimensional electron' fields and three '8-dimensional neutrino' fields.

However, our 4-dimensional space-time admits only 2-dimensional curls. There are two types of 2-dimensional curls; these are the 2-dimensional space-time curl like the conventional electric field and the 2-dimensional Euclidean curl like the the conventional magnetic field. It is these types of curls we will extract from the 8-dimensional algebras. We opine that all other types of curl cannot be manifest in our 4-dimensional space-time.

The reader might be aware that conventional particle physics posits two types of quark, the up quark and the down quark. The up quark has twice the electric charge of the down quark.

The 8-dimensional algebras:
We reiterate, there are seven sets of 128 non-commutative 8-dimensional Clifford algebras that derive from the finite group $C_2 \times C_2 \times C_2$.[106] Each of these seven sets is comprised of the same 128 algebras. The 128 algebras of a particular set are of three algebraically distinct types which are:

$$P_{5,1} = +1 \sim \begin{cases} 16 & \textit{off} & Cl_{0,3} = 1 + \sqrt{+1} + 6\sqrt{-1} \\ 48 & \textit{off} & Cl_{2,1} = 1 + 5\sqrt{+1} + 2\sqrt{-1} \end{cases} \tag{24.1}$$

$$P_{5,1} = -1 \sim \left\{ 64 \quad \textit{off} \quad Cl_{3,0} \cong Cl_{1,3} = 1 + 3\sqrt{+1} + 4\sqrt{-1}\,_{Non-com} \right\}$$

The algebras in each of these sets of 16, 48, and 64, algebras are algebraically isomorphic, but the individual algebras differ from each other in that that do not necessarily have the same commutation relations. Thus, instead of two 4-dimensional algebras, we have three 8-dimensional algebras. Instead of two quaternion algebras and six A_3 algebras, we have 16, 48, & 64 8-dimensional algebras.

---

[106] There is also one set of 128 commutative algebras.

First, a look at some details:

Within a particular set of 128 algebras, above, we have given each algebra an arbitrary case number; for example, in this book, the sixteen $Cl_{0,3} = 1 + \sqrt{+1} + 6\sqrt{-1}$ algebras are the algebras with the case numbers: 2, 4, 10, 12, 18, 20, 26, 28, 34, 36, 42, 44, 50, 52, 58, and 60. The case numbers are peculiar to a particular set of 128 algebras. The case numbers are peculiar to a particular author; the case numbers are peculiar to the particular general algebraic matrix form, and the case numbers are peculiar to the computer program which allocates them to each algebra. The case numbers are imposed arbitrarily.

Within a set of 128 non-commutative 8-dimensional Clifford algebras that derive from the finite group $C_2 \times C_2 \times C_2$, the algebras can be collated together into 32 sets of four algebras, $\{A, B, C, D\}$, based on the commutation relations of those algebras. We call such a set of four algebras an algebra marriage.

Notation:
We use the notation:

$$\text{Algebra Marriage} \sim \begin{Bmatrix} \{A, B\} \\ \{C, D\} \end{Bmatrix} \tag{24.2}$$

To indicate an algebra marriage of four 8-dimensional Clifford algebras:

a) The algebra A has the same commutation relations as the algebra B.
b) The algebra C has the same commutation relations as the algebra D.
c) The commutation relations of the A & B algebras are the opposite of the commutation relations of the C & D algebras.
d) The algebraic matrix forms of the algebras A & C commute.
e) The algebraic matrix forms of the algebras B & D commute.
f) The sum $A - B - C + D = 0$ is the zero matrix.

Algebra marriages:
The algebra marriages of the sixteen $Cl_{0,3} = 1 + \sqrt{+1} + 6\sqrt{-1}$ algebras are:

$$\begin{Bmatrix} \{2, 20\} \\ \{42, 60\} \end{Bmatrix} \quad \begin{Bmatrix} \{4, 18\} \\ \{44, 58\} \end{Bmatrix} \quad \begin{Bmatrix} \{10, 28\} \\ \{34, 52\} \end{Bmatrix} \quad \begin{Bmatrix} \{12, 26\} \\ \{36, 50\} \end{Bmatrix} \tag{24.3}$$

The 8-dimensional potentials:
The potentials used to form the E-fields and the B-fields are based upon conjugated algebraic matrix form similar to the case 2 potential:

$$\Phi_{Case\ 2} = \begin{bmatrix} \phi_a & -A_b & -A_c & -A_d & \phi_e & -A_f & -A_g & -A_h \\ A_b & \phi_a & -A_d & A_c & -A_f & -\phi_e & -A_h & A_g \\ A_c & A_d & \phi_a & -A_b & -A_g & A_h & -\phi_e & -A_f \\ A_d & -A_c & A_b & \phi_a & -A_h & -A_g & A_f & -\phi_e \\ \phi_e & A_f & A_g & A_h & \phi_a & A_b & A_c & A_d \\ A_f & -\phi_e & -A_h & A_g & -A_b & \phi_a & -A_d & A_c \\ A_g & A_h & -\phi_e & -A_f & -A_c & A_d & \phi_a & -A_b \\ A_h & -A_g & A_f & -\phi_e & -A_d & -A_c & A_b & \phi_a \end{bmatrix} \qquad (24.4)$$

The E-fields of the sixteen $Cl_{0,3}$ algebras:

The E-fields of these sixteen algebras are given below. The variables in the differentials are the same in each algebra; only the signs differ, and so we show only the signs.

	E-Fields of the $Cl_{0,3} = 1 + \sqrt{+1} + 6\sqrt{-1}$ algebras – page 1		
Marriage	Case	$E_{[1,1]}$	$E_{[1,5]}$
$\left\{ \begin{array}{l} \{2,20\} \\ \{42,60\} \end{array} \right\}$	2	$\dfrac{\partial \phi_a}{\partial t} - \dfrac{\partial A_b}{\partial b} - \dfrac{\partial A_c}{\partial c} - \dfrac{\partial A_d}{\partial d} +$ $\dfrac{\partial \phi_e}{\partial e} - \dfrac{\partial A_f}{\partial f} - \dfrac{\partial A_g}{\partial g} - \dfrac{\partial A_h}{\partial h}$ $(+,-,-,-,+,-,-,-)$	$\dfrac{\partial \phi_a}{\partial e} + \dfrac{\partial A_b}{\partial f} + \dfrac{\partial A_c}{\partial g} + \dfrac{\partial A_d}{\partial h} +$ $\dfrac{\partial \phi_e}{\partial t} + \dfrac{\partial A_f}{\partial b} + \dfrac{\partial A_g}{\partial c} + \dfrac{\partial A_h}{\partial d}$ $(+,+,+,+,+,+,+,+)$
	20	$(+,-,-,-,+,-,-,-)$	$(+,-,-,-,+,-,-,-)$
	42	$(+,-,-,-,+,-,-,-)$	$(+,+,+,+,+,+,+,+)$
	60	$(+,-,-,-,+,-,-,-)$	$(+,-,-,-,+,-,-,-)$
$\left\{ \begin{array}{l} \{4,18\} \\ \{44,58\} \end{array} \right\}$	4	$(+,-,-,-,+,-,-,-)$	$(+,+,-,-,+,+,-,-)$
	18	$(+,-,-,-,+,-,-,-)$	$(+,-,+,+,+,-,+,+)$
	44	$(+,-,-,-,+,-,-,-)$	$(+,+,-,-,+,+,-,-)$
	58	$(+,-,-,-,+,-,-,-)$	$(+,-,+,+,+,-,+,+)$
$\left\{ \begin{array}{l} \{10,28\} \\ \{34,52\} \end{array} \right\}$	10	$(+,-,-,-,+,-,-,-)$	$(+,+,+,-,+,+,+,-)$
	28	$(+,-,-,-,+,-,-,-)$	$(+,-,-,+,+,-,-,+)$
	34	$(+,-,-,-,+,-,-,-)$	$(+,+,+,-,+,+,+,-)$
	52	$(+,-,-,-,+,-,-,-)$	$(+,-,-,+,+,-,-,+)$
$\left\{ \begin{array}{l} \{12,26\} \\ \{36,50\} \end{array} \right\}$	12	$(+,-,-,-,+,-,-,-)$	$(+,+,-,+,+,+,-,+)$
	26	$(+,-,-,-,+,-,-,-)$	$(+,-,+,-,+,-,+,-)$
	36	$(+,-,-,-,+,-,-,-)$	$(+,+,-,+,+,+,-,+)$
	50	$(+,-,-,-,+,-,-,-)$	$(+,-,+,-,+,-,+,-)$

Totals		$16(+,-,-,+,-,-,-)$	$16(+,0,0,0,+,0,0,0)$
Total		$\dfrac{\partial \phi_a}{\partial t} - \dfrac{\partial A_b}{\partial b} - \dfrac{\partial A_c}{\partial c} - \dfrac{\partial A_d}{\partial d} +$ $\dfrac{\partial \phi_e}{\partial e} - \dfrac{\partial A_f}{\partial f} - \dfrac{\partial A_g}{\partial g} - \dfrac{\partial A_h}{\partial h}$	$\dfrac{\partial \phi_a}{\partial e} + \dfrac{\partial \phi_e}{\partial t}$

E-Fields of the $Cl_{0,3} = 1 + \sqrt{+1} + 6\sqrt{-1}$ algebras – page 2				
Marriage	Case	$E_{[1,2]}$	$E_{[1,3]}$	$E_{[1,4]}$
$\left\{ \begin{array}{l} \{2,20\} \\ \{42,60\} \end{array} \right\}$	2	$-\dfrac{\partial \phi_a}{\partial b} - \dfrac{\partial A_b}{\partial t} + \dfrac{\partial \phi_e}{\partial f} + \dfrac{\partial A_f}{\partial e}$ $(-,-,+,+)$	$-\dfrac{\partial \phi_a}{\partial c} - \dfrac{\partial A_c}{\partial t} + \dfrac{\partial \phi_e}{\partial g} + \dfrac{\partial A_g}{\partial e}$ $(-,-,+,+)$	$-\dfrac{\partial \phi_a}{\partial d} - \dfrac{\partial A_d}{\partial t} + \dfrac{\partial \phi_e}{\partial h} + \dfrac{\partial A_h}{\partial e}$ $(-,-,+,+)$
	20	$(-,-,-,-)$	$(-,-,-,-)$	$(-,-,-,-)$
	42	$(-,-,+,+)$	$(-,-,+,+)$	$(-,-,+,+)$
	60	$(-,-,-,-)$	$(-,-,-,-)$	$(-,-,-,-)$
$\left\{ \begin{array}{l} \{4,18\} \\ \{44,58\} \end{array} \right\}$	4	$(-,-,+,+)$	$(-,-,-,-)$	$(-,-,-,-)$
	18	$(-,-,-,-)$	$(-,-,+,+)$	$(-,-,+,+)$
	44	$(-,-,+,+)$	$(-,-,-,-)$	$(-,-,-,-)$
	58	$(-,-,-,-)$	$(-,-,+,+)$	$(-,-,+,+)$
$\left\{ \begin{array}{l} \{10,28\} \\ \{34,52\} \end{array} \right\}$	10	$(-,-,+,+)$	$(-,-,+,+)$	$(-,-,-,-)$
	28	$(-,-,-,-)$	$(-,-,-,-)$	$(-,-,+,+)$
	34	$(-,-,+,+)$	$(-,-,+,+)$	$(-,-,-,-)$
	52	$(-,-,-,-)$	$(-,-,-,-)$	$(-,-,+,+)$
$\left\{ \begin{array}{l} \{12,26\} \\ \{36,50\} \end{array} \right\}$	12	$(-,-,+,+)$	$(-,-,-,-)$	$(-,-,+,+)$
	26	$(-,-,-,-)$	$(-,-,+,+)$	$(-,-,-,-)$
	36	$(\ ,\ ,+,+)$	$(-,-,-,-)$	$(-,-,+,+)$
	50	$(-,-,-,-)$	$(-,-,+,+)$	$(-,-,-,-)$
Totals		$16(-,-,0,0)$	$16(-,-,0,0)$	$16(-,-,0,0)$
Total		$-\dfrac{\partial \phi_a}{\partial b} - \dfrac{\partial A_b}{\partial t}$	$-\dfrac{\partial \phi_a}{\partial c} - \dfrac{\partial A_c}{\partial t}$	$-\dfrac{\partial \phi_a}{\partial d} - \dfrac{\partial A_d}{\partial t}$

E-Fields of the $Cl_{0,3} = 1 + \sqrt{+1} + 6\sqrt{-1}$ algebras – page 3				
Marriage	Case	$E_{[1,6]}$	$E_{[1,7]}$	$E_{[1,8]}$

	2	$-\dfrac{\partial\phi_a}{\partial f}-\dfrac{\partial A_f}{\partial t}+\dfrac{\partial\phi_e}{\partial b}+\dfrac{\partial A_b}{\partial e}$	$-\dfrac{\partial\phi_a}{\partial g}-\dfrac{\partial A_g}{\partial t}+\dfrac{\partial\phi_e}{\partial c}+\dfrac{\partial A_c}{\partial e}$	$-\dfrac{\partial\phi_a}{\partial h}-\dfrac{\partial A_h}{\partial t}+\dfrac{\partial\phi_e}{\partial d}+\dfrac{\partial A_d}{\partial e}$
$\left\{\begin{matrix}\{2,20\}\\\{42,60\}\end{matrix}\right\}$		$(-,-,+,+)$	$(-,-,+,+)$	$(-,-,+,+)$
	20	$(-,-,-,-)$	$(-,-,-,-)$	$(-,-,-,-)$
	42	$(-,-,+,+)$	$(-,-,+,+)$	$(-,-,+,+)$
	60	$(-,-,-,-)$	$(-,-,-,-)$	$(-,-,-,-)$
$\left\{\begin{matrix}\{4,18\}\\\{44,58\}\end{matrix}\right\}$	4	$(-,-,+,+)$	$(-,-,-,-)$	$(-,-,-,-)$
	18	$(-,-,-,-)$	$(-,-,+,+)$	$(-,-,+,+)$
	44	$(-,-,+,+)$	$(-,-,-,-)$	$(-,-,-,-)$
	58	$(-,-,-,-)$	$(-,-,+,+)$	$(-,-,+,+)$
$\left\{\begin{matrix}\{10,28\}\\\{34,52\}\end{matrix}\right\}$	10	$(-,-,+,+)$	$(-,-,+,+)$	$(-,-,-,-)$
	28	$(-,-,-,-)$	$(-,-,-,-)$	$(-,-,+,+)$
	34	$(-,-,+,+)$	$(-,-,+,+)$	$(-,-,-,-)$
	52	$(-,-,-,-)$	$(-,-,-,-)$	$(-,-,+,+)$
$\left\{\begin{matrix}\{12,26\}\\\{36,50\}\end{matrix}\right\}$	12	$(-,-,+,+)$	$(-,-,-,-)$	$(-,-,+,+)$
	26	$(-,-,-,-)$	$(-,-,+,+)$	$(-,-,-,-)$
	36	$(-,-,+,+)$	$(-,-,-,-)$	$(-,-,+,+)$
	50	$(-,-,-,-)$	$(-,-,+,+)$	$(-,-,-,-)$
Totals		$16(-,-,0,0)$	$16(-,-,0,0)$	$16(-,-,0,0)$
Total		$-\dfrac{\partial\phi_a}{\partial f}-\dfrac{\partial A_f}{\partial t}$	$-\dfrac{\partial\phi_a}{\partial g}-\dfrac{\partial A_g}{\partial t}$	$-\dfrac{\partial\phi_a}{\partial h}-\dfrac{\partial A_h}{\partial t}$

Looking at the above lists of E-fields of the $Cl_{0,3}=1+\sqrt{+1}+6\sqrt{-1}$ algebras, we will take the E-fields of

the algebra marriage $\left\{\begin{matrix}\{2,20\}\\\{42,60\}\end{matrix}\right\}$. Ignoring the 4, we add the E-fields of these four algebras to get:

The $Cl_{0,3}=1+\sqrt{+1}+6\sqrt{-1}$ marriage divergence, $E_{[1,1]}^{\left\{\begin{matrix}\{2,20\}\\\{42,60\}\end{matrix}\right\}}$ is:

$$Div\left(\left\{\begin{matrix}\{2,20\}\\\{42,60\}\end{matrix}\right\}\right)=E_{[1,1]}^{\left\{\begin{matrix}\{2,20\}\\\{42,60\}\end{matrix}\right\}}=\frac{\partial\phi_a}{\partial t}-\frac{\partial A_b}{\partial b}-\frac{\partial A_c}{\partial c}-\frac{\partial A_d}{\partial d}$$
$$+\frac{\partial\phi_e}{\partial e}-\frac{\partial A_f}{\partial f}-\frac{\partial A_g}{\partial g}-\frac{\partial A_h}{\partial h}$$

$$(24.5)$$

This is of the form of the divergences of two 4-dimensional electric fields.

We also have the $Cl_{0,3}=1+\sqrt{+1}+6\sqrt{-1}$ marriage E-field:

$$E_{[1,5]}^{\left\{\substack{\{2,20\} \\ \{42,60\}}\right\}} = \frac{\partial \phi_a}{\partial e} + \frac{\partial \phi_e}{\partial t} \tag{24.6}$$

We also have the $Cl_{0,3} = 1 + \sqrt{+1} + 6\sqrt{-1}$ marriage E-field:

$$E_{[1,2]}^{\left\{\substack{\{2,20\} \\ \{42,60\}}\right\}} = -\frac{\partial \phi_a}{\partial b} - \frac{\partial A_b}{\partial t} \qquad\qquad E_{[1,6]}^{\left\{\substack{\{2,20\} \\ \{42,60\}}\right\}} = -\frac{\partial \phi_a}{\partial f} - \frac{\partial A_f}{\partial t}$$

$$E_{[1,3]}^{\left\{\substack{\{2,20\} \\ \{42,60\}}\right\}} = -\frac{\partial \phi_a}{\partial c} - \frac{\partial A_c}{\partial t} \qquad\qquad E_{[1,7]}^{\left\{\substack{\{2,20\} \\ \{42,60\}}\right\}} = -\frac{\partial \phi_a}{\partial g} - \frac{\partial A_g}{\partial t} \tag{24.7}$$

$$E_{[1,4]}^{\left\{\substack{\{2,20\} \\ \{42,60\}}\right\}} = -\frac{\partial \phi_a}{\partial d} - \frac{\partial A_d}{\partial t} \qquad\qquad E_{[1,8]}^{\left\{\substack{\{2,20\} \\ \{42,60\}}\right\}} = -\frac{\partial \phi_a}{\partial h} - \frac{\partial A_h}{\partial t}$$

We have the three spatial parts two electric fields. Inspection of the table above shows we have the same results for every $Cl_{0,3} = 1 + \sqrt{+1} + 6\sqrt{-1}$ algebra marriage.

Taking all the $Cl_{0,3} = 1 + \sqrt{+1} + 6\sqrt{-1}$ algebras together gives:

$$Div = E_{[1,1]}^{Total\ Cl_{0,3}} = 16 \left(\begin{array}{l} \dfrac{\partial \phi_a}{\partial t} - \dfrac{\partial A_b}{\partial b} - \dfrac{\partial A_c}{\partial c} - \dfrac{\partial A_d}{\partial d} \\[2mm] + \dfrac{\partial \phi_e}{\partial e} - \dfrac{\partial A_f}{\partial f} - \dfrac{\partial A_g}{\partial g} - \dfrac{\partial A_h}{\partial h} \end{array} \right) \tag{24.8}$$

$$E_{[1,5]}^{Total\ Cl_{0,3}} = 16 \left(\frac{\partial \phi_a}{\partial e} + \frac{\partial \phi_e}{\partial t} \right) \tag{24.9}$$

$$E_{[1,2]}^{Total\ Cl_{0,3}} = 16 \left(-\frac{\partial \phi_a}{\partial b} - \frac{\partial A_b}{\partial t} \right) \qquad\qquad E_{[1,6]}^{Total\ Cl_{0,3}} = 16 \left(-\frac{\partial \phi_a}{\partial f} - \frac{\partial A_f}{\partial t} \right)$$

$$E_{[1,3]}^{Total\ Cl_{0,3}} = 16 \left(-\frac{\partial \phi_a}{\partial c} - \frac{\partial A_c}{\partial t} \right) \qquad\qquad E_{[1,7]}^{Total\ Cl_{0,3}} = 16 \left(-\frac{\partial \phi_a}{\partial g} - \frac{\partial A_g}{\partial t} \right) \tag{24.10}$$

$$E_{[1,4]}^{Total\ Cl_{0,3}} = 16 \left(-\frac{\partial \phi_a}{\partial d} - \frac{\partial A_d}{\partial t} \right) \qquad\qquad E_{[1,8]}^{Total\ Cl_{0,3}} = 16 \left(-\frac{\partial \phi_a}{\partial h} - \frac{\partial A_h}{\partial t} \right)$$

We have the emergent $Cl_{0,3} = 1 + \sqrt{+1} + 6\sqrt{-1}$ E-field:

$$\begin{bmatrix}
Div & E^{Total\ Cl_{0,3}}_{[1,2]} & E^{Total\ Cl_{0,3}}_{[1,3]} & E^{Total\ Cl_{0,3}}_{[1,4]} & E^{Total\ Cl_{0,3}}_{[1,5]} & E^{Total\ Cl_{0,3}}_{[1,6]} & E^{Total\ Cl_{0,3}}_{[1,7]} & E^{Total\ Cl_{0,3}}_{[1,8]} \\[4pt]
-E^{Total\ Cl_{0,3}}_{[1,2]} & Div & 0 & 0 & \dfrac{\partial \phi_e}{\partial b}+\dfrac{\partial A_b}{\partial e} & -\dfrac{\partial A_b}{\partial f}-\dfrac{\partial A_f}{\partial b} & 0 & 0 \\[10pt]
-E^{Total\ Cl_{0,3}}_{[1,3]} & 0 & Div & 0 & \dfrac{\partial \phi_e}{\partial c}+\dfrac{\partial A_c}{\partial e} & 0 & -\dfrac{\partial A_c}{\partial g}-\dfrac{\partial A_g}{\partial c} & 0 \\[10pt]
-E^{Total\ Cl_{0,3}}_{[1,4]} & 0 & 0 & Div & \dfrac{\partial \phi_e}{\partial d}+\dfrac{\partial A_d}{\partial e} & 0 & 0 & -\dfrac{\partial A_d}{\partial h}-\dfrac{\partial A_h}{\partial d} \\[10pt]
E^{Total\ Cl_{0,3}}_{[1,5]} & -E_{[2,5]} & -E_{[3,5]} & -E_{[4,5]} & Div & -E_{[6,5]} & -E_{[7,5]} & -E_{[8,5]} \\[10pt]
-E^{Total\ Cl_{0,3}}_{[1,6]} & E_{[2,6]} & 0 & 0 & \dfrac{\partial \phi_e}{\partial f}+\dfrac{\partial A_f}{\partial e} & Div & 0 & 0 \\[10pt]
-E^{Total\ Cl_{0,3}}_{[1,7]} & 0 & E_{[3,7]} & 0 & \dfrac{\partial \phi_e}{\partial g}+\dfrac{\partial A_g}{\partial e} & 0 & Div & 0 \\[10pt]
-E^{Total\ Cl_{0,3}}_{[1,8]} & 0 & 0 & E_{[4,8]} & \dfrac{\partial \phi_e}{\partial h}+\dfrac{\partial A_h}{\partial e} & 0 & 0 & Div
\end{bmatrix}$$

$$(24.11)$$

We see that this is similar to the quaternion E-field but that we have some new terms. If we ignore the 4×4 top right-hand corner and the 4×4 bottom left-hand corner, we have:

$$\begin{bmatrix}
Div & E^{Total\ Cl_{0,3}}_{[1,2]} & E^{Total\ Cl_{0,3}}_{[1,3]} & E^{Total\ Cl_{0,3}}_{[1,4]} & 0 & 0 & 0 & 0 \\[6pt]
-E^{Total\ Cl_{0,3}}_{[1,2]} & Div & 0 & 0 & 0 & 0 & 0 & 0 \\[6pt]
-E^{Total\ Cl_{0,3}}_{[1,3]} & 0 & Div & 0 & 0 & 0 & 0 & 0 \\[6pt]
-E^{Total\ Cl_{0,3}}_{[1,4]} & 0 & 0 & Div & 0 & 0 & 0 & 0 \\[6pt]
0 & 0 & 0 & 0 & Div & -E_{[6,5]} & -E_{[7,5]} & -E_{[8,5]} \\[10pt]
0 & 0 & 0 & 0 & \dfrac{\partial \phi_e}{\partial f}+\dfrac{\partial A_f}{\partial e} & Div & 0 & 0 \\[10pt]
0 & 0 & 0 & 0 & \dfrac{\partial \phi_e}{\partial g}+\dfrac{\partial A_g}{\partial e} & 0 & Div & 0 \\[10pt]
0 & 0 & 0 & 0 & \dfrac{\partial \phi_e}{\partial h}+\dfrac{\partial A_h}{\partial e} & 0 & 0 & Div
\end{bmatrix}\quad(24.12)$$

This is of the form of two oppositely charged electron fields if we allow the e variable to play the same role as the time variable.

We found anti-matter in the quaternions in the form of a potential with a negative real variable. Within the 8-dimensional algebras, we cannot have a negative real variable and maintain the algebraic structure of the algebra[107], and so we might expect no 8-dimensional anti-matter. Looking at the above, (24.12),

---

[107] Remember, we must take the exponential of the 8-dimensional algebraic matrix form to produce the division algebra.

contrary to our expectations, it seems that an '8-dimensional electron' field can come as a 4-dimensional 'electron' and a 4-dimensional 'positron' tied together if we allow the e variable to be a time variable.

The role of the e variable is intriguing. This is an imaginary variable, and so it can take both positive and negative values. It also seems to 'think' it is a time variable. The CPT theorem states that the combination of symmetries charge conjugation, parity, and time-reversal is never violated. Since CP invariance is violated, time-reversal symmetry must also be violated, but this cannot happen unless time can flow in reverse. Perhaps the e variable is involved in this violation of time-reversal symmetry.

The E-fields of the forty-eight $Cl_{2,1}$ algebras:

The algebra marriages of the forty-eight $Cl_{2,1} = 1 + 5\sqrt{+1} + 2\sqrt{-1}$ algebras are:

$$\begin{Bmatrix} \{6,24\} \\ \{46,64\} \end{Bmatrix}, \begin{Bmatrix} \{8,22\} \\ \{48,62\} \end{Bmatrix}, \begin{Bmatrix} \{14,32\} \\ \{38,56\} \end{Bmatrix}, \begin{Bmatrix} \{16,30\} \\ \{40,54\} \end{Bmatrix}, \begin{Bmatrix} \{66,84\} \\ \{106,124\} \end{Bmatrix}, \begin{Bmatrix} \{68,82\} \\ \{108,122\} \end{Bmatrix}$$

$$\begin{Bmatrix} \{70,88\} \\ \{110,128\} \end{Bmatrix}, \begin{Bmatrix} \{72,86\} \\ \{112,126\} \end{Bmatrix}, \begin{Bmatrix} \{74,92\} \\ \{98,116\} \end{Bmatrix}, \begin{Bmatrix} \{76,90\} \\ \{100,114\} \end{Bmatrix}, \begin{Bmatrix} \{78,96\} \\ \{102,120\} \end{Bmatrix}, \begin{Bmatrix} \{80,94\} \\ \{104,118\} \end{Bmatrix}$$

$$(24.13)$$

E-Fields of the $Cl_{2,1} = 1 + 5\sqrt{+1} + 2\sqrt{-1}$ algebras – page 1			
Marriage	Case	$E_{[1,1]}$	$E_{[1,5]}$
$\begin{Bmatrix} \{6,24\} \\ \{46,64\} \end{Bmatrix}$	6	$\dfrac{\partial \phi_a}{\partial t} - \dfrac{\partial A_b}{\partial b} - \dfrac{\partial A_c}{\partial c} - \dfrac{\partial A_d}{\partial d}$ $\dfrac{\partial \phi_e}{\partial e} - \dfrac{\partial A_f}{\partial f} - \dfrac{\partial A_g}{\partial g} - \dfrac{\partial A_h}{\partial h}$ $(+,-,-,-,+,-,-,-)$	$\dfrac{\partial \phi_a}{\partial e} + \dfrac{\partial A_b}{\partial f} + \dfrac{\partial A_c}{\partial g} + \dfrac{\partial A_d}{\partial h}$ $+\dfrac{\partial \phi_e}{\partial t} + \dfrac{\partial A_f}{\partial b} + \dfrac{\partial A_g}{\partial c} + \dfrac{\partial A_h}{\partial d}$ $(+,+,+,+,+,+,+,+)$
	24	$(+,-,-,-,+,-,-,-)$	$(+,-,-,-,+,-,-,-)$
	46	$(+,-,-,-,+,-,-,-)$	$(+,+,+,+,+,+,+,+)$
	64	$(+,-,-,-,+,-,-,-)$	$(+,-,-,-,+,-,-,-)$

E-Fields of the $Cl_{2,1} = 1 + 5\sqrt{+1} + 2\sqrt{-1}$ algebras – page 2				
Marriage	Case	$E_{[1,2]}$	$E_{[1,3]}$	$E_{[1,4]}$

{6,24} {46,64}	6	$-\dfrac{\partial\phi_a}{\partial b} - \dfrac{\partial A_b}{\partial t} + \dfrac{\partial\phi_e}{\partial f} + \dfrac{\partial A_f}{\partial e}$ (−,−,+,+)	$\dfrac{\partial\phi_a}{\partial c} - \dfrac{\partial A_c}{\partial t} - \dfrac{\partial\phi_e}{\partial g} + \dfrac{\partial A_g}{\partial e}$ (+,−,−,+)	$\dfrac{\partial\phi_a}{\partial d} - \dfrac{\partial A_d}{\partial t} - \dfrac{\partial\phi_e}{\partial h} + \dfrac{\partial A_h}{\partial e}$ (+,−,−,+)
	24	(−,−,−,−)	(+,−,+,−)	(+,−,+,−)
	46	(−,−,+,+)	(+,−,−,+)	(+,−,−,+)
	64	(−,−,−,−)	(+,−,+,−)	(+,−,+,−)

E-Fields of the algebras $Cl_{2,1} = 1+5\sqrt{+1}+2\sqrt{-1}$ – page 3				
Marriage	Case	$E_{[1,6]}$	$E_{[1,7]}$	$E_{[1,8]}$
{6,24} {46,64}	6	$-\dfrac{\partial\phi_a}{\partial f} - \dfrac{\partial A_f}{\partial t} + \dfrac{\partial\phi_e}{\partial b} + \dfrac{\partial A_b}{\partial e}$ (−,−,+,+)	$\dfrac{\partial\phi_a}{\partial g} - \dfrac{\partial A_g}{\partial t} - \dfrac{\partial\phi_e}{\partial c} + \dfrac{\partial A_c}{\partial e}$ (+,−,−,+)	$\dfrac{\partial\phi_a}{\partial h} - \dfrac{\partial A_h}{\partial t} - \dfrac{\partial\phi_e}{\partial d} + \dfrac{\partial A_d}{\partial e}$ (+,−,−,+)
	24	(−,−,−,−)	(+,−,+,−)	(+,−,+,−)
	46	(−,−,+,+)	(+,−,−,+)	(+,−,−,+)
	64	(−,−,−,−)	(+,−,+,−)	(+,−,+,−)

The $Cl_{2,1} = 1+5\sqrt{+1}+2\sqrt{-1}$ marriage divergence, $E_{[1,1]}^{\left\{\substack{\{6,24\}\\\{46,64\}}\right\}}$ is:

$$Div\left(\left\{\substack{\{6,24\}\\\{46,64\}}\right\}\right) = E_{[1,1]}^{\left\{\substack{\{6,24\}\\\{46,64\}}\right\}} = \frac{\partial\phi_a}{\partial t} - \frac{\partial A_b}{\partial b} - \frac{\partial A_c}{\partial c} - \frac{\partial A_d}{\partial d}$$
$$+ \frac{\partial\phi_e}{\partial e} - \frac{\partial A_f}{\partial f} - \frac{\partial A_g}{\partial g} - \frac{\partial A_h}{\partial h} \tag{24.14}$$

All the $Cl_{2,1} = 1+5\sqrt{+1}+2\sqrt{-1}$ marriages have the same $E_{[1,1]}$. This is of the form of the divergences of two electric fields.

We also have the $Cl_{2,1} = 1+5\sqrt{+1}+2\sqrt{-1}$ marriage E-field:

$$E_{[1,5]}^{\left\{\substack{\{6,24\}\\\{46,64\}}\right\}} = \frac{\partial\phi_a}{\partial e} + \frac{\partial\phi_e}{\partial t} \tag{24.15}$$

All the $Cl_{2,1} = 1+5\sqrt{+1}+2\sqrt{-1}$ marriages have the same $E_{[1,5]}$.

We also have the $Cl_{2,1} = 1 + 5\sqrt{+1} + 2\sqrt{-1} \begin{Bmatrix} \{6,24\} \\ \{46,64\} \end{Bmatrix}$ marriage E-field:

$$E_{[1,2]}^{\begin{Bmatrix} \{6,24\} \\ \{46,64\} \end{Bmatrix}} = -\frac{\partial\phi_a}{\partial b} - \frac{\partial A_b}{\partial t} \qquad\qquad E_{[1,6]}^{\begin{Bmatrix} \{6,24\} \\ \{46,64\} \end{Bmatrix}} = -\frac{\partial\phi_a}{\partial f} - \frac{\partial A_f}{\partial t}$$

$$E_{[1,3]}^{\begin{Bmatrix} \{6,24\} \\ \{46,64\} \end{Bmatrix}} = +\frac{\partial\phi_a}{\partial c} - \frac{\partial A_c}{\partial t} \qquad\qquad E_{[1,7]}^{\begin{Bmatrix} \{6,24\} \\ \{46,64\} \end{Bmatrix}} = +\frac{\partial\phi_a}{\partial g} - \frac{\partial A_g}{\partial t} \qquad\qquad (24.16)$$

$$E_{[1,4]}^{\begin{Bmatrix} \{6,24\} \\ \{46,64\} \end{Bmatrix}} = +\frac{\partial\phi_a}{\partial d} - \frac{\partial A_d}{\partial t} \qquad\qquad E_{[1,8]}^{\begin{Bmatrix} \{6,24\} \\ \{46,64\} \end{Bmatrix}} = +\frac{\partial\phi_a}{\partial h} - \frac{\partial A_h}{\partial t}$$

We note that this differs from (24.7) by the plus sign in $E_{[1,2]}^{\begin{Bmatrix} \{6,24\} \\ \{46,64\} \end{Bmatrix}}$ & $E_{[1,6]}^{\begin{Bmatrix} \{6,24\} \\ \{46,64\} \end{Bmatrix}}$.

The 'unwanted plus sign' in the E-fields in (24.16) occurs in different places throughout the other eleven $Cl_{2,1} = 1 + 5\sqrt{+1} + 2\sqrt{-1}$ marriages.

Another example is:

E-Fields of the $Cl_{2,1} = 1 + 5\sqrt{+1} + 2\sqrt{-1}$ algebras – page 1			
Marriage	Case	$E_{[1,1]}$	$E_{[1,5]}$
$\begin{Bmatrix} \{8,22\} \\ \{48,62\} \end{Bmatrix}$	8	$\frac{\partial\phi_a}{\partial t} - \frac{\partial A_b}{\partial b} - \frac{\partial A_c}{\partial c} - \frac{\partial A_d}{\partial d}$ $\frac{\partial\phi_e}{\partial e} - \frac{\partial A_f}{\partial f} - \frac{\partial A_g}{\partial g} - \frac{\partial A_h}{\partial h}$ $(+,-,-,-,+,-,-,-)$	$\frac{\partial\phi_a}{\partial e} + \frac{\partial A_b}{\partial f} + \frac{\partial A_c}{\partial g} + \frac{\partial A_d}{\partial h}$ $+\frac{\partial\phi_e}{\partial t} + \frac{\partial A_f}{\partial b} + \frac{\partial A_g}{\partial c} + \frac{\partial A_h}{\partial d}$ $(+,+,-,-,+,+,-,-)$
	22	$(+,-,-,-,+,-,-,-)$	$(+,-,+,+,+,-,+,+)$
	48	$(+,-,-,-,+,-,-,-)$	$(+,+,-,-,+,+,-,-)$
	62	$(+,-,-,-,+,-,-,-)$	$(+,-,+,+,+,-,+,+)$

E-Fields of the $Cl_{2,1} = 1 + 5\sqrt{+1} + 2\sqrt{-1}$ algebras – page 2				
Marriage	Case	$E_{[1,2]}$	$E_{[1,3]}$	$E_{[1,4]}$

{8,22} {48,62}	8	$-\dfrac{\partial \phi_a}{\partial b}-\dfrac{\partial A_b}{\partial t}+\dfrac{\partial \phi_e}{\partial f}+\dfrac{\partial A_f}{\partial e}$ $(-,-,+,+)$	$\dfrac{\partial \phi_a}{\partial c}-\dfrac{\partial A_c}{\partial t}-\dfrac{\partial \phi_e}{\partial g}+\dfrac{\partial A_g}{\partial e}$ $(+,-,+,-)$	$\dfrac{\partial \phi_a}{\partial d}-\dfrac{\partial A_d}{\partial t}-\dfrac{\partial \phi_e}{\partial h}+\dfrac{\partial A_h}{\partial e}$ $(+,-,+,-)$
	22	$(-,-,-,-)$	$(+,-,-,+)$	$(+,-,-,+)$
	48	$(-,-,+,+)$	$(+,-,+,-)$	$(+,-,+,-)$
	62	$(-,-,-,-)$	$(+,-,-,+)$	$(+,-,-,+)$

E-Fields of the algebras $Cl_{2,1}=1+5\sqrt{+1}+2\sqrt{-1}$ – page 3				
Marriage	Case	$E_{[1,6]}$	$E_{[1,7]}$	$E_{[1,8]}$
{8,22} {48,62}	8	$-\dfrac{\partial \phi_a}{\partial f}-\dfrac{\partial A_f}{\partial t}+\dfrac{\partial \phi_e}{\partial b}+\dfrac{\partial A_b}{\partial e}$ $(-,-,+,+)$	$\dfrac{\partial \phi_a}{\partial g}-\dfrac{\partial A_g}{\partial t}-\dfrac{\partial \phi_e}{\partial c}+\dfrac{\partial A_c}{\partial e}$ $(+,-,+,-)$	$\dfrac{\partial \phi_a}{\partial h}-\dfrac{\partial A_h}{\partial t}-\dfrac{\partial \phi_e}{\partial d}+\dfrac{\partial A_d}{\partial e}$ $(+,-,+,-)$
	22	$(-,-,-,-)$	$(+,-,-,+)$	$(+,-,-,+)$
	48	$(-,-,+,+)$	$(+,-,+,-)$	$(+,-,+,-)$
	62	$(-,-,-,-)$	$(+,-,-,+)$	$(+,-,-,+)$

The $Cl_{2,1}=1+5\sqrt{+1}+2\sqrt{-1}$ marriage divergence, $E_{[1,1]}^{\left\{\substack{\{8,22\}\\ \{48,62\}}\right\}}$ is:

$$Div\left(\left\{\substack{\{8,22\}\\ \{48,62\}}\right\}\right)=E_{[1,1]}^{\left\{\substack{\{8,22\}\\ \{48,62\}}\right\}}=\frac{\partial \phi_a}{\partial t}-\frac{\partial A_b}{\partial b}-\frac{\partial A_c}{\partial c}-\frac{\partial A_d}{\partial d}$$
$$+\frac{\partial \phi_e}{\partial e}-\frac{\partial A_f}{\partial f}-\frac{\partial A_g}{\partial g}-\frac{\partial A_h}{\partial h} \tag{24.17}$$

All the $Cl_{2,1}=1+5\sqrt{+1}+2\sqrt{-1}$ marriages have the same $E_{[1,1]}$. This is of the form of the divergences of two electric fields.

We also have the $Cl_{2,1}=1+5\sqrt{+1}+2\sqrt{-1}$ marriage E-field:

$$E_{[1,5]}^{\left\{\substack{\{8,22\}\\ \{48,62\}}\right\}}=\frac{\partial \phi_a}{\partial e}+\frac{\partial \phi_e}{\partial t} \tag{24.18}$$

We also have the $Cl_{2,1}=1+5\sqrt{+1}+2\sqrt{-1}$ $\left\{\substack{\{8,22\}\\ \{48,62\}}\right\}$ marriage E-field:

$$E_{[1,2]}^{\left\{\substack{\{8,22\}\\\{48,62\}}\right\}} = -\frac{\partial \phi_a}{\partial b} - \frac{\partial A_b}{\partial t} \qquad\qquad E_{[1,6]}^{\left\{\substack{\{8,22\}\\\{48,62\}}\right\}} = -\frac{\partial \phi_a}{\partial f} - \frac{\partial A_f}{\partial t}$$

$$E_{[1,3]}^{\left\{\substack{\{8,22\}\\\{48,62\}}\right\}} = +\frac{\partial \phi_a}{\partial c} - \frac{\partial A_c}{\partial t} \qquad\qquad E_{[1,7]}^{\left\{\substack{\{8,22\}\\\{48,62\}}\right\}} = +\frac{\partial \phi_a}{\partial g} - \frac{\partial A_g}{\partial t} \qquad (24.19)$$

$$E_{[1,4]}^{\left\{\substack{\{8,22\}\\\{48,62\}}\right\}} = +\frac{\partial \phi_a}{\partial d} - \frac{\partial A_d}{\partial t} \qquad\qquad E_{[1,8]}^{\left\{\substack{\{8,22\}\\\{48,62\}}\right\}} = +\frac{\partial \phi_a}{\partial h} - \frac{\partial A_h}{\partial t}$$

The E-fields of these algebras are basically of the form of (24.19), but the double minus signs can change position; for example, consider the marriage $\left\{\substack{\{68,82\}\\\{108,122\}}\right\}$. We have:

$$E_{[1,2]}^{\left\{\substack{\{68,82\}\\\{108,122\}}\right\}} = +\frac{\partial \phi_a}{\partial b} - \frac{\partial A_b}{\partial t} \qquad\qquad E_{[1,6]}^{\left\{\substack{\{68,82\}\\\{108,122\}}\right\}} = +\frac{\partial \phi_a}{\partial f} - \frac{\partial A_f}{\partial t}$$

$$E_{[1,3]}^{\left\{\substack{\{68,82\}\\\{108,122\}}\right\}} = -\frac{\partial \phi_a}{\partial c} - \frac{\partial A_c}{\partial t} \qquad\qquad E_{[1,7]}^{\left\{\substack{\{68,82\}\\\{108,122\}}\right\}} = -\frac{\partial \phi_a}{\partial g} - \frac{\partial A_g}{\partial t} \qquad (24.20)$$

$$E_{[1,4]}^{\left\{\substack{\{68,82\}\\\{108,122\}}\right\}} = +\frac{\partial \phi_a}{\partial d} - \frac{\partial A_d}{\partial t} \qquad\qquad E_{[1,8]}^{\left\{\substack{\{68,82\}\\\{108,122\}}\right\}} = +\frac{\partial \phi_a}{\partial h} - \frac{\partial A_h}{\partial t}$$

Or the marriage $\left\{\substack{\{70,88\}\\\{110,128\}}\right\}$:

$$E_{[1,2]}^{\left\{\substack{\{70,88\}\\\{110,128\}}\right\}} = +\frac{\partial \phi_a}{\partial b} - \frac{\partial A_b}{\partial t} \qquad\qquad E_{[1,6]}^{\left\{\substack{\{70,88\}\\\{110,128\}}\right\}} = +\frac{\partial \phi_a}{\partial f} - \frac{\partial A_f}{\partial t}$$

$$E_{[1,3]}^{\left\{\substack{\{70,88\}\\\{110,128\}}\right\}} = +\frac{\partial \phi_a}{\partial c} - \frac{\partial A_c}{\partial t} \qquad\qquad E_{[1,7]}^{\left\{\substack{\{70,88\}\\\{110,128\}}\right\}} = +\frac{\partial \phi_a}{\partial g} - \frac{\partial A_g}{\partial t} \qquad (24.21)$$

$$E_{[1,4]}^{\left\{\substack{\{70,88\}\\\{110,128\}}\right\}} = -\frac{\partial \phi_a}{\partial d} - \frac{\partial A_d}{\partial t} \qquad\qquad E_{[1,8]}^{\left\{\substack{\{70,88\}\\\{110,128\}}\right\}} = -\frac{\partial \phi_a}{\partial h} - \frac{\partial A_h}{\partial t}$$

The E-field total for all forty-eight $Cl_{2,1} = 1 + 5\sqrt{+1} + 2\sqrt{-1}$ algebras, sixteen algebra marriages, is:

$$E_{[1,1]}^{Total\ Cl_{2,1}} = 48 \left(\begin{array}{l} \dfrac{\partial \phi_a}{\partial t} - \dfrac{\partial A_b}{\partial b} - \dfrac{\partial A_c}{\partial c} - \dfrac{\partial A_d}{\partial d} \\[2mm] +\dfrac{\partial \phi_e}{\partial e} - \dfrac{\partial A_f}{\partial f} - \dfrac{\partial A_g}{\partial g} - \dfrac{\partial A_h}{\partial h} \end{array} \right) \qquad (24.22)$$

$$E_{[1,5]}^{Total\ Cl_{2,1}} = 48 \left(\frac{\partial \phi_a}{\partial e} + \frac{\partial \phi_e}{\partial t} \right) \qquad (24.23)$$

137

$$E_{[1,2]}^{Total\ Cl_{2,1}} = +16\frac{\partial\phi_a}{\partial b} - 48\frac{\partial A_b}{\partial t} \qquad\qquad E_{[1,6]}^{Total\ Cl_{2,1}} = +16\frac{\partial\phi_a}{\partial f} - 48\frac{\partial A_f}{\partial t}$$

$$E_{[1,3]}^{Total\ Cl_{2,1}} = +16\frac{\partial\phi_a}{\partial c} - 48\frac{\partial A_c}{\partial t} \qquad\qquad E_{[1,7]}^{Total\ Cl_{2,1}} = +16\frac{\partial\phi_a}{\partial g} - 48\frac{\partial A_g}{\partial t} \qquad (24.24)$$

$$E_{[1,4]}^{Total\ Cl_{2,1}} = +16\frac{\partial\phi_a}{\partial d} - 48\frac{\partial A_d}{\partial t} \qquad\qquad E_{[1,8]}^{Total\ Cl_{2,1}} = +16\frac{\partial\phi_a}{\partial h} - 48\frac{\partial A_h}{\partial t}$$

We have the emergent $Cl_{2,1} = 1 + 5\sqrt{+1} + 2\sqrt{-1}$ E-field:

$$
\begin{bmatrix}
Div & E_{[1,2]}^{Total\ Cl_{2,1}} & E_{[1,3]}^{Total\ Cl_{2,1}} & E_{[1,4]}^{Total\ Cl_{2,1}} & E_{[1,5]}^{Total\ Cl_{2,1}} & E_{[1,6]}^{Total\ Cl_{2,1}} & E_{[1,7]}^{Total\ Cl_{2,1}} & E_{[1,8]}^{Total\ Cl_{2,1}} \\[2mm]
-E_{[1,2]}^{Total\ Cl_{2,1}} & Div & 0 & 0 & 48\frac{\partial\phi_e}{\partial b}-16\frac{\partial A_b}{\partial e} & 48\left(-\frac{\partial A_b}{\partial f}-\frac{\partial A_f}{\partial b}\right) & 0 & 0 \\[2mm]
-E_{[1,3]}^{Total\ Cl_{2,1}} & 0 & Div & 0 & 48\frac{\partial\phi_e}{\partial c}-16\frac{\partial A_c}{\partial e} & 0 & 48\left(-\frac{\partial A_c}{\partial g}-\frac{\partial A_g}{\partial c}\right) & 0 \\[2mm]
-E_{[1,4]}^{Total\ Cl_{2,1}} & 0 & 0 & Div & 48\frac{\partial\phi_e}{\partial d}-16\frac{\partial A_d}{\partial e} & 0 & 0 & 48\left(-\frac{\partial A_d}{\partial h}-\frac{\partial A_h}{\partial d}\right) \\[2mm]
E_{[1,5]}^{Total\ Cl_{2,1}} & -E_{[2,5]} & -E_{[3,5]} & -E_{[4,5]} & Div & -E_{[6,5]} & -E_{[7,5]} & -E_{[8,5]} \\[2mm]
-E_{[1,6]}^{Total\ Cl_{2,1}} & E_{[2,6]} & 0 & 0 & 48\frac{\partial\phi_e}{\partial f}-16\frac{\partial A_f}{\partial e} & Div & 0 & 0 \\[2mm]
-E_{[1,7]}^{Total\ Cl_{2,1}} & 0 & E_{[3,7]} & 0 & 48\frac{\partial\phi_e}{\partial g}-16\frac{\partial A_g}{\partial e} & 0 & Div & 0 \\[2mm]
-E_{[1,8]}^{Total\ Cl_{2,1}} & 0 & 0 & E_{[4,8]} & 48\frac{\partial\phi_e}{\partial h}-16\frac{\partial A_h}{\partial e} & 0 & 0 & Div
\end{bmatrix}
$$

$$(24.25)$$

This is basically the same form as the above emergent E-field, (24.11) except that it is unbalanced.

The E-fields of the sixty-four Cl$_{3,0}$ algebras:

The algebra marriages of the sixty-four $Cl_{3,0} \cong Cl_{1,3} = 1 + 3\sqrt{+1} + 4\sqrt{-1}_{Non-com}$ algebras are:

$$
\left\{\begin{matrix}\{1,19\}\\\{41,59\}\end{matrix}\right\},\
\left\{\begin{matrix}\{3,17\}\\\{43,57\}\end{matrix}\right\},\
\left\{\begin{matrix}\{5,23\}\\45,63\end{matrix}\right\},\
\left\{\begin{matrix}\{7,21\}\\\{47,61\}\end{matrix}\right\},\
\left\{\begin{matrix}\{9,27\}\\\{33,51\}\end{matrix}\right\},\
\left\{\begin{matrix}\{11,25\}\\\{35,49\}\end{matrix}\right\}
$$

$$
\left\{\begin{matrix}\{13,31\}\\\{37,55\}\end{matrix}\right\},\
\left\{\begin{matrix}\{15,29\}\\\{39,53\}\end{matrix}\right\},\
\left\{\begin{matrix}\{65,83\}\\\{105,123\}\end{matrix}\right\},\
\left\{\begin{matrix}\{67,81\}\\\{107.121\}\end{matrix}\right\},\
\left\{\begin{matrix}\{69,87\}\\\{109,127\}\end{matrix}\right\} \qquad (24.26)
$$

$$
\left\{\begin{matrix}\{71,85\}\\\{111,125\}\end{matrix}\right\},\
\left\{\begin{matrix}\{73,91\}\\\{97,115\}\end{matrix}\right\},\
\left\{\begin{matrix}\{75,89\}\\\{99,113\}\end{matrix}\right\},\
\left\{\begin{matrix}\{77,95\}\\\{101,119\}\end{matrix}\right\},\
\left\{\begin{matrix}\{79,93\}\\\{103,117\}\end{matrix}\right\}
$$

E-Fields of the $Cl_{3,0} \cong Cl_{1,3} = 1 + 3\sqrt{+1} + 4\sqrt{-1}_{Non-com}$ algebras – page 1

Marriage	Case	$E_{[1,1]}$	$E_{[1,5]}$
$\begin{cases} \{1,19\} \\ \{41,59\} \end{cases}$	1	$\dfrac{\partial \phi_a}{\partial t} - \dfrac{\partial A_b}{\partial b} - \dfrac{\partial A_c}{\partial c} - \dfrac{\partial A_d}{\partial d} + \dfrac{\partial \phi_e}{\partial e} - \dfrac{\partial A_f}{\partial f} - \dfrac{\partial A_g}{\partial g} - \dfrac{\partial A_h}{\partial h}$ $(+,-,-,-,+,-,-,-)$	$-\dfrac{\partial \phi_a}{\partial e} - \dfrac{\partial A_b}{\partial f} - \dfrac{\partial A_c}{\partial g} - \dfrac{\partial A_d}{\partial h} + \dfrac{\partial \phi_e}{\partial t} + \dfrac{\partial A_f}{\partial b} + \dfrac{\partial A_g}{\partial c} + \dfrac{\partial A_h}{\partial d}$ $(-,-,-,-,+,+,+,+)$
	19	$(-,-,-,-,-,-,-,-)$	$(-,+,+,+,+,-,-,-)$
	41	$(-,-,-,-,-,-,-,-)$	$(-,-,-,-,+,+,+,+)$
	59	$(-,-,-,-,-,-,-,-)$	$(-,+,+,+,+,-,-,-)$

E-Fields of the $Cl_{3,0} \cong Cl_{1,3} = 1 + 3\sqrt{+1} + 4\sqrt{-1}_{Non-com}$ algebras – page 2

Marriage	Case	$E_{[1,2]}$	$E_{[1,3]}$	$E_{[1,4]}$
$\begin{cases} \{1,19\} \\ \{41,59\} \end{cases}$	1	$-\dfrac{\partial \phi_a}{\partial b} - \dfrac{\partial A_b}{\partial t} + \dfrac{\partial \phi_e}{\partial f} + \dfrac{\partial A_f}{\partial e}$ $(-,-,+,+)$	$-\dfrac{\partial \phi_a}{\partial c} - \dfrac{\partial A_c}{\partial t} + \dfrac{\partial \phi_e}{\partial g} + \dfrac{\partial A_g}{\partial e}$ $(-,-,+,+)$	$-\dfrac{\partial \phi_a}{\partial d} - \dfrac{\partial A_d}{\partial t} + \dfrac{\partial \phi_e}{\partial h} - \dfrac{\partial A_h}{\partial e}$ $(-,-,+,+)$
	19	$(-,-,-,-)$	$(-,-,-,-)$	$(-,-,-,-)$
	41	$(-,-,+,+)$	$(-,-,+,+)$	$(-,-,+,+)$
	59	$(-,-,-,-)$	$(-,-,-,-)$	$(-,-,-,-)$

E-Fields of the algebras $Cl_{3,0} \cong Cl_{1,3} = 1 + 3\sqrt{+1} + 4\sqrt{-1}_{Non-com}$ – page 3

Marriage	Case	$E_{[1,6]}$	$E_{[1,7]}$	$E_{[1,8]}$
$\begin{cases} \{1,19\} \\ \{41,59\} \end{cases}$	1	$\dfrac{\partial \phi_a}{\partial f} - \dfrac{\partial A_f}{\partial t} + \dfrac{\partial \phi_e}{\partial b} - \dfrac{\partial A_b}{\partial e}$ $(+,-,+,-)$	$\dfrac{\partial \phi_a}{\partial g} - \dfrac{\partial A_g}{\partial t} + \dfrac{\partial \phi_e}{\partial c} - \dfrac{\partial A_c}{\partial e}$ $(+,-,+,-)$	$\dfrac{\partial \phi_a}{\partial h} - \dfrac{\partial A_h}{\partial t} + \dfrac{\partial \phi_e}{\partial d} - \dfrac{\partial A_d}{\partial e}$ $(+,-,+,-)$
	19	$(+,-,-,+)$	$(+,-,-,+)$	$(+,-,-,+)$
	41	$(+,-,+,-)$	$(+,-,+,-)$	$(+,-,+,-)$
	59	$(+,-,-,+)$	$(+,-,-,+)$	$(+,-,-,+)$

The $Cl_{3,0} \cong Cl_{1,3} = 1 + 3\sqrt{+1} + 4\sqrt{-1}_{Non-com}$ marriage divergence, $E_{[1,1]}^{\left\{\begin{smallmatrix}\{1,19\}\\\{41,59\}\end{smallmatrix}\right\}}$ is:

$$Div\left(\left\{\begin{matrix}\{1,19\}\\\{41,59\}\end{matrix}\right\}\right) = E_{[1,1]}^{\left\{\begin{smallmatrix}\{1,19\}\\\{41,59\}\end{smallmatrix}\right\}} = \frac{\partial\phi_a}{\partial t} - \frac{\partial A_b}{\partial b} - \frac{\partial A_c}{\partial c} - \frac{\partial A_d}{\partial d}$$
$$+ \frac{\partial\phi_e}{\partial e} - \frac{\partial A_f}{\partial f} - \frac{\partial A_g}{\partial g} - \frac{\partial A_h}{\partial h}$$

(24.27)

All the $Cl_{3,0} \cong Cl_{1,3} = 1 + 3\sqrt{+1} + 4\sqrt{-1}_{Non-com}$ marriages have the same $E_{[1,1]}$.

We also have the $Cl_{3,0} \cong Cl_{1,3} = 1 + \sqrt{+1} + 6\sqrt{-1}_{non-com}$ marriage E-field:

$$E_{[1,5]}^{\left\{\begin{smallmatrix}\{1,19\}\\\{41,59\}\end{smallmatrix}\right\}} = -\frac{\partial\phi_a}{\partial e} + \frac{\partial\phi_e}{\partial t}$$

(24.28)

All the $Cl_{3,0} \cong Cl_{1,3} = 1 + 3\sqrt{+1} + 4\sqrt{-1}_{Non-com}$ marriages have the same $E_{[1,5]}$.

We also have the $Cl_{3,0} \cong Cl_{1,3} = 1 + 3\sqrt{+1} + 4\sqrt{-1}_{Non-com}$ marriage E-field:

$$E_{[1,2]}^{\left\{\begin{smallmatrix}\{1,19\}\\\{41,59\}\end{smallmatrix}\right\}} = -\frac{\partial\phi_a}{\partial b} - \frac{\partial A_b}{\partial t} \qquad E_{[1,6]}^{\left\{\begin{smallmatrix}\{1,19\}\\\{41,59\}\end{smallmatrix}\right\}} = +\frac{\partial\phi_a}{\partial f} - \frac{\partial A_f}{\partial t}$$

$$E_{[1,3]}^{\left\{\begin{smallmatrix}\{1,19\}\\\{41,59\}\end{smallmatrix}\right\}} = -\frac{\partial\phi_a}{\partial c} - \frac{\partial A_c}{\partial t} \qquad E_{[1,2]}^{\left\{\begin{smallmatrix}\{1,19\}\\\{41,59\}\end{smallmatrix}\right\}} = +\frac{\partial\phi_a}{\partial g} - \frac{\partial A_g}{\partial t}$$

$$E_{[1,4]}^{\left\{\begin{smallmatrix}\{1,19\}\\\{41,59\}\end{smallmatrix}\right\}} = -\frac{\partial\phi_a}{\partial d} - \frac{\partial A_d}{\partial t} \qquad E_{[1,2]}^{\left\{\begin{smallmatrix}\{1,19\}\\\{41,59\}\end{smallmatrix}\right\}} = +\frac{\partial\phi_a}{\partial h} - \frac{\partial A_h}{\partial t}$$

(24.29)

This leads to a total $Cl_{3,0} \cong Cl_{1,3} = 1 + 3\sqrt{+1} + 4\sqrt{-1}_{Non-com}$ E-field:

$$E_{[1,1]}^{Total\ Cl_{3,0}} = 64\left(\begin{matrix}\frac{\partial\phi_a}{\partial t} - \frac{\partial A_b}{\partial b} - \frac{\partial A_c}{\partial c} - \frac{\partial A_d}{\partial d}\\+\frac{\partial\phi_e}{\partial e} - \frac{\partial A_f}{\partial f} - \frac{\partial A_g}{\partial g} - \frac{\partial A_h}{\partial h}\end{matrix}\right)$$

$$E_{[1,5]}^{Total\ Cl_{3,0}} = 64\left(-\frac{\partial\phi_a}{\partial e} + \frac{\partial\phi_e}{\partial t}\right)$$

(24.30)

And:

$$E_{[1,2]}^{Total\ Cl_{3,0}} = -64\frac{\partial A_b}{\partial t} \qquad\qquad E_{[1,6]}^{Total\ Cl_{3,0}} = -64\frac{\partial A_f}{\partial t}$$

$$E_{[1,3]}^{Total\ Cl_{3,0}} = -64\frac{\partial A_c}{\partial t} \qquad\qquad E_{[1,2]}^{Total\ Cl_{3,0}} = -64\frac{\partial A_g}{\partial t} \qquad\qquad (24.31)$$

$$E_{[1,4]}^{Total\ Cl_{3,0}} = -64\frac{\partial A_d}{\partial t} \qquad\qquad E_{[1,2]}^{Total\ Cl_{3,0}} = -64\frac{\partial A_h}{\partial t}$$

The emergent $Cl_{3,0} \cong Cl_{1,3} = 1 + 3\sqrt{+1} + 4\sqrt{-1}\,_{Non-com}$ E-field is:

$$64 \begin{bmatrix}
Div & -\dfrac{\partial A_b}{\partial t} & -\dfrac{\partial A_c}{\partial t} & -\dfrac{\partial A_d}{\partial t} & \dfrac{\partial \phi_e}{\partial t} - \dfrac{\partial \phi_a}{\partial e} & -\dfrac{\partial A_f}{\partial t} & -\dfrac{\partial A_g}{\partial t} & -\dfrac{\partial A_h}{\partial t} \\[2.2ex]
\dfrac{\partial A_b}{\partial t} & Div & 0 & 0 & \dfrac{\partial \phi_e}{\partial b} & \dfrac{\partial A_b}{\partial f} - \dfrac{\partial A_f}{\partial b} & 0 & 0 \\[2.2ex]
\dfrac{\partial A_c}{\partial t} & 0 & Div & 0 & \dfrac{\partial \phi_e}{\partial c} & 0 & \dfrac{\partial A_c}{\partial g} - \dfrac{\partial A_g}{\partial c} & 0 \\[2.2ex]
\dfrac{\partial A_d}{\partial t} & 0 & 0 & Div & \dfrac{\partial \phi_e}{\partial e} & 0 & 0 & \dfrac{\partial A_d}{\partial h} - \dfrac{\partial A_h}{\partial d} \\[2.2ex]
-\dfrac{\partial \phi_e}{\partial t} + \dfrac{\partial \phi_a}{\partial e} & -\dfrac{\partial A_b}{\partial e} & -\dfrac{\partial A_c}{\partial e} & -\dfrac{\partial A_d}{\partial e} & Div & -\dfrac{\partial A_f}{\partial e} & -\dfrac{\partial A_g}{\partial e} & -\dfrac{\partial A_h}{\partial e} \\[2.2ex]
\dfrac{\partial A_f}{\partial t} & -E_{[2,6]} & 0 & 0 & \dfrac{\partial \phi_e}{\partial f} & Div & 0 & 0 \\[2.2ex]
\dfrac{\partial A_g}{\partial t} & 0 & -E_{[3,7]} & 0 & \dfrac{\partial \phi_e}{\partial g} & 0 & Div & 0 \\[2.2ex]
\dfrac{\partial A_h}{\partial t} & 0 & 0 & -E_{[4,8]} & \dfrac{\partial \phi_e}{\partial h} & 0 & 0 & Div
\end{bmatrix}$$

$$(24.32)$$

Bringing it all together:

We see that every one of the 128 8-dimensional algebras has a divergence which twice matches the divergence of our 4-dimensional space-time if we allow that the e variable is a kind of time variable:

$$Div = E_{[1,1]} = +\frac{\partial \phi_a}{\partial t} - \frac{\partial A_b}{\partial b} - \frac{\partial A_c}{\partial c} - \frac{\partial A_d}{\partial d}$$
$$+\frac{\partial \phi_e}{\partial e} - \frac{\partial A_f}{\partial f} - \frac{\partial A_g}{\partial g} - \frac{\partial A_h}{\partial h} \qquad (24.33)$$

We have two types of $E_{[1,5]}$ field components. We have:

$$P_{5,1} = +1 \sim \begin{cases} E_{[1,5]} = 16\left(\dfrac{\partial \phi_a}{\partial e} + \dfrac{\partial \phi_e}{\partial t}\right) \\[3mm] E_{[1,5]} = 48\left(\dfrac{\partial \phi_a}{\partial e} + \dfrac{\partial \phi_e}{\partial t}\right) \end{cases}$$

(24.34)

$$P_{5,1} = -1 \sim E_{[1,5]} = 64\left(-\dfrac{\partial \phi_a}{\partial e} + \dfrac{\partial \phi_e}{\partial t}\right)$$

We also have:

$$E_{[1,2]}^{Total\ Cl_{0,3}} = 16\left(-\dfrac{\partial \phi_a}{\partial b} - \dfrac{\partial A_b}{\partial t}\right) \qquad E_{[1,6]}^{Total\ Cl_{0,3}} = 16\left(-\dfrac{\partial \phi_a}{\partial f} - \dfrac{\partial A_f}{\partial t}\right)$$

$$E_{[1,3]}^{Total\ Cl_{0,3}} = 16\left(-\dfrac{\partial \phi_a}{\partial c} - \dfrac{\partial A_c}{\partial t}\right) \qquad E_{[1,7]}^{Total\ Cl_{0,3}} = 16\left(-\dfrac{\partial \phi_a}{\partial g} - \dfrac{\partial A_g}{\partial t}\right) \qquad (24.35)$$

$$E_{[1,4]}^{Total\ Cl_{0,3}} = 16\left(-\dfrac{\partial \phi_a}{\partial d} - \dfrac{\partial A_d}{\partial t}\right) \qquad E_{[1,8]}^{Total\ Cl_{0,3}} = 16\left(-\dfrac{\partial \phi_a}{\partial h} - \dfrac{\partial A_h}{\partial t}\right)$$

And:

$$E_{[1,2]}^{Total\ Cl_{2,1}} = +16\dfrac{\partial \phi_a}{\partial b} - 48\dfrac{\partial A_b}{\partial t} \qquad E_{[1,6]}^{Total\ Cl_{2,1}} = +16\dfrac{\partial \phi_a}{\partial f} - 48\dfrac{\partial A_f}{\partial t}$$

$$E_{[1,3]}^{Total\ Cl_{2,1}} = +16\dfrac{\partial \phi_a}{\partial c} - 48\dfrac{\partial A_c}{\partial t} \qquad E_{[1,7]}^{Total\ Cl_{2,1}} = +16\dfrac{\partial \phi_a}{\partial g} - 48\dfrac{\partial A_g}{\partial t} \qquad (24.36)$$

$$E_{[1,4]}^{Total\ Cl_{2,1}} = +16\dfrac{\partial \phi_a}{\partial d} - 48\dfrac{\partial A_d}{\partial t} \qquad E_{[1,8]}^{Total\ Cl_{2,1}} = +16\dfrac{\partial \phi_a}{\partial h} - 48\dfrac{\partial A_h}{\partial t}$$

And:

$$E_{[1,2]}^{Total\ Cl_{3,0}} = -64\dfrac{\partial A_b}{\partial t} \qquad E_{[1,6]}^{Total\ Cl_{3,0}} = -64\dfrac{\partial A_f}{\partial t}$$

$$E_{[1,3]}^{Total\ Cl_{3,0}} = -64\dfrac{\partial A_c}{\partial t} \qquad E_{[1,2]}^{Total\ Cl_{3,0}} = -64\dfrac{\partial A_g}{\partial t} \qquad (24.37)$$

$$E_{[1,4]}^{Total\ Cl_{3,0}} = -64\dfrac{\partial A_d}{\partial t} \qquad E_{[1,2]}^{Total\ Cl_{3,0}} = -64\dfrac{\partial A_h}{\partial t}$$

Curls in our 4-dimensional space-time:

Our 4-dimensional space-time is comprised of six 2-dimensional spinor spaces; there are three $\mathbb{C}$ spinor spaces and three $\mathbb{S}$ spinor spaces. These two types of 2-dimensional spaces admit only two types of 2-dimensional curl which are a space-time curl and a spatial curl:

$$Curl_{space-time} = \dfrac{\partial \phi}{\partial x} + \dfrac{\partial A_x}{\partial t} \qquad \& \qquad Curl_{Spatial} = \dfrac{\partial A_x}{\partial y} - \dfrac{\partial A_y}{\partial x} \qquad (24.38)$$

We therefore expect to see only these types of curls in our 4-dimensional space-time. When we look at (24.34), we see these types of curls do not appear in our 4-dimensional space-time. Looking at (24.35), we see that these curls do appear in our 4-dimensional space-time. Looking at (24.36), we see these types of curls do not appear in our 4-dimensional space-time. Looking at (24.37), we see these types of curls do not appear in our 4-dimensional space-time.

At this point, it seems that only the 'electric field' type of curl, (24.35), appears in our 4-dimensional space-time and that only the $Cl_{0,3} = 1 + \sqrt{+1} + 6\sqrt{-1}$ algebra is manifest in our 4-dimensional space-time. Things are not that simple.

Picking Out the E-Fields

Well! We have a plethora of E-field components. What are we to do with them all?

What are all these fields?

When we take a quaternion potential and differentiate it non-commutatively, we see the electron field and the neutrino field fall out quite simply. The 8-dimensional case is much more complicated. We have an important unanswered question:

1) How does an 8-dimensional field manifest itself in our 4-dimensional space-time?

Perhaps we see only the 4-dimensional sub-algebras, or does the 8-dimensional space 'fold' into a 4-dimensional space. Perhaps these 8-dimensional fields do not manifest themselves in our 4-dimensional space-time and we will never see the equivalent of a '8-dimensional electron' in our 4-dimensional space-time. The astute reader will know that quarks are fermions just as the electron is a fermion and that we never see a quark. Thus, our '8-dimensional electrons' might be quarks and the E-fields might be quark fields.

In short, we do not know how an 8-dimensional field will manifest itself in our 4-dimensional space-time; however, we can make a guess.

A guess:

There are two types of curl in our 4-dimensional space-time. We have the wholly spatial curl of the 2-dimensional complex numbers, $\mathbb{C}$, which is of the form of a plus sign and a minus sign:

$$Curl_x = \frac{\partial A_z}{\partial y} - \frac{\partial A_y}{\partial z}, \qquad Curl_y = \frac{\partial A_x}{\partial z} - \frac{\partial A_z}{\partial x}, \qquad Curl_z = \frac{\partial A_y}{\partial x} - \frac{\partial A_x}{\partial y} \qquad (25.1)$$

We also have the space-time curl of the 2-dimensional hyperbolic complex numbers, $\mathbb{S}$, which is of the form of both signs the same:

$$E_x = -\frac{\partial \phi}{\partial x} - \frac{\partial A_x}{\partial t}, \qquad E_y = -\frac{\partial \phi}{\partial y} - \frac{\partial A_y}{\partial t}, \qquad E_z = -\frac{\partial \phi}{\partial z} - \frac{\partial A_z}{\partial t} \qquad (25.2)$$

We guess that only such fields can be manifest in our 4-dimensional space-time.

We will pick out from the 8-dimensional E-fields only those fields which are of the form (25.2). The remaining E-fields we will ignore on the basis that these other fields cannot fit into our 4-dimensional space-time. We are interested in only the totals of these fields for each of the three types of 8-dimensional algebra, but we will list the details for the perusal of the reader.

List of the electric type fields:

We take each marriage, and we form the sum of the E-fields of the four algebras of that marriage. We seek fields which are the same form as the spatial components of the familiar electric field in our 4-dimensional space-time. We will leave blank cells in the table that do not have the same form as the spatial components of the familiar electric field in our 4-dimensional space-time. We have:

The sixteen and forty-eight $P_{5,1} = +1$ algebras:

Marriage	$E_{[1,2]}$	$E_{[1,3]}$	$E_{[1,4]}$	$E_{[1,6]}$	$E_{[1,7]}$	$E_{[1,8]}$
$\begin{Bmatrix} \{2,20\} \\ \{42,60\} \end{Bmatrix}$	$-\dfrac{\partial \phi_a}{\partial b} - \dfrac{\partial A_b}{\partial t}$	$-\dfrac{\partial \phi_a}{\partial c} - \dfrac{\partial A_c}{\partial t}$	$-\dfrac{\partial \phi_a}{\partial d} - \dfrac{\partial A_d}{\partial t}$	$-\dfrac{\partial \phi_a}{\partial f} - \dfrac{\partial A_f}{\partial t}$	$-\dfrac{\partial \phi_a}{\partial g} - \dfrac{\partial A_g}{\partial t}$	$-\dfrac{\partial \phi_a}{\partial h} - \dfrac{\partial A_h}{\partial t}$
$\begin{Bmatrix} \{4,18\} \\ \{44,58\} \end{Bmatrix}$	$-\dfrac{\partial \phi_a}{\partial b} - \dfrac{\partial A_b}{\partial t}$	$-\dfrac{\partial \phi_a}{\partial c} - \dfrac{\partial A_c}{\partial t}$	$-\dfrac{\partial \phi_a}{\partial d} - \dfrac{\partial A_d}{\partial t}$	$-\dfrac{\partial \phi_a}{\partial f} - \dfrac{\partial A_f}{\partial t}$	$-\dfrac{\partial \phi_a}{\partial g} - \dfrac{\partial A_g}{\partial t}$	$-\dfrac{\partial \phi_a}{\partial h} - \dfrac{\partial A_h}{\partial t}$
$\begin{Bmatrix} \{10,28\} \\ \{34,52\} \end{Bmatrix}$	$-\dfrac{\partial \phi_a}{\partial b} - \dfrac{\partial A_b}{\partial t}$	$-\dfrac{\partial \phi_a}{\partial c} - \dfrac{\partial A_c}{\partial t}$	$-\dfrac{\partial \phi_a}{\partial d} - \dfrac{\partial A_d}{\partial t}$	$-\dfrac{\partial \phi_a}{\partial f} - \dfrac{\partial A_f}{\partial t}$	$-\dfrac{\partial \phi_a}{\partial g} - \dfrac{\partial A_g}{\partial t}$	$-\dfrac{\partial \phi_a}{\partial h} - \dfrac{\partial A_h}{\partial t}$
$\begin{Bmatrix} \{12,26\} \\ \{36,50\} \end{Bmatrix}$	$-\dfrac{\partial \phi_a}{\partial b} - \dfrac{\partial A_b}{\partial t}$	$-\dfrac{\partial \phi_a}{\partial c} - \dfrac{\partial A_c}{\partial t}$	$-\dfrac{\partial \phi_a}{\partial d} - \dfrac{\partial A_d}{\partial t}$	$-\dfrac{\partial \phi_a}{\partial f} - \dfrac{\partial A_f}{\partial t}$	$-\dfrac{\partial \phi_a}{\partial g} - \dfrac{\partial A_g}{\partial t}$	$-\dfrac{\partial \phi_a}{\partial h} - \dfrac{\partial A_h}{\partial t}$
-------	-------	-------	-------	-------	-------	-------
$\begin{Bmatrix} \{6,24\} \\ \{46,64\} \end{Bmatrix}$	$-\dfrac{\partial \phi_a}{\partial b} - \dfrac{\partial A_b}{\partial t}$			$-\dfrac{\partial \phi_a}{\partial f} - \dfrac{\partial A_f}{\partial t}$		
$\begin{Bmatrix} \{8,22\} \\ \{48,62\} \end{Bmatrix}$	$-\dfrac{\partial \phi_a}{\partial b} - \dfrac{\partial A_b}{\partial t}$			$-\dfrac{\partial \phi_a}{\partial f} - \dfrac{\partial A_f}{\partial t}$		
$\begin{Bmatrix} \{14,32\} \\ \{38,56\} \end{Bmatrix}$	$-\dfrac{\partial \phi_a}{\partial b} - \dfrac{\partial A_b}{\partial t}$			$-\dfrac{\partial \phi_a}{\partial f} - \dfrac{\partial A_f}{\partial t}$		
$\begin{Bmatrix} \{16,30\} \\ \{40,54\} \end{Bmatrix}$	$-\dfrac{\partial \phi_a}{\partial b} - \dfrac{\partial A_b}{\partial t}$			$-\dfrac{\partial \phi_a}{\partial f} - \dfrac{\partial A_f}{\partial t}$		
$\begin{Bmatrix} \{66,84\} \\ \{106,124\} \end{Bmatrix}$		$-\dfrac{\partial \phi_a}{\partial c} - \dfrac{\partial A_c}{\partial t}$			$-\dfrac{\partial \phi_a}{\partial g} - \dfrac{\partial A_g}{\partial t}$	
$\begin{Bmatrix} \{68,82\} \\ \{108,122\} \end{Bmatrix}$		$-\dfrac{\partial \phi_a}{\partial c} - \dfrac{\partial A_c}{\partial t}$			$-\dfrac{\partial \phi_a}{\partial g} - \dfrac{\partial A_g}{\partial t}$	
$\begin{Bmatrix} \{70,88\} \\ \{110,128\} \end{Bmatrix}$			$-\dfrac{\partial \phi_a}{\partial d} - \dfrac{\partial A_d}{\partial t}$			$-\dfrac{\partial \phi_a}{\partial h} - \dfrac{\partial A_h}{\partial t}$
$\begin{Bmatrix} \{72,86\} \\ \{112,126\} \end{Bmatrix}$			$-\dfrac{\partial \phi_a}{\partial d} - \dfrac{\partial A_d}{\partial t}$			$-\dfrac{\partial \phi_a}{\partial h} - \dfrac{\partial A_h}{\partial t}$
$\begin{Bmatrix} \{74,92\} \\ \{98,116\} \end{Bmatrix}$		$-\dfrac{\partial \phi_a}{\partial c} - \dfrac{\partial A_c}{\partial t}$			$-\dfrac{\partial \phi_a}{\partial g} - \dfrac{\partial A_g}{\partial t}$	

Marriage	$E_{[1,2]}$	$E_{[1,3]}$	$E_{[1,4]}$	$E_{[1,6]}$	$E_{[1,7]}$	$E_{[1,8]}$
$\left\{\begin{array}{l}\{76,90\}\\\{100,114\}\end{array}\right\}$		$-\dfrac{\partial\phi_a}{\partial c}-\dfrac{\partial A_c}{\partial t}$			$-\dfrac{\partial\phi_a}{\partial g}-\dfrac{\partial A_g}{\partial t}$	
$\left\{\begin{array}{l}\{78,96\}\\\{102,120\}\end{array}\right\}$			$-\dfrac{\partial\phi_a}{\partial d}-\dfrac{\partial A_d}{\partial t}$			$-\dfrac{\partial\phi_a}{\partial h}-\dfrac{\partial A_h}{\partial t}$
$\left\{\begin{array}{l}\{80,94\}\\\{104,118\}\end{array}\right\}$			$-\dfrac{\partial\phi_a}{\partial d}-\dfrac{\partial A_d}{\partial t}$			$-\dfrac{\partial\phi_a}{\partial h}-\dfrac{\partial A_h}{\partial t}$

And the sixty-four algebras:

Marriage	$E_{[1,2]}$	$E_{[1,3]}$	$E_{[1,4]}$	$E_{[1,6]}$	$E_{[1,7]}$	$E_{[1,8]}$
$\left\{\begin{array}{l}\{1,19\}\\\{41,59\}\end{array}\right\}$	$-\dfrac{\partial\phi_a}{\partial b}-\dfrac{\partial A_b}{\partial t}$	$-\dfrac{\partial\phi_a}{\partial c}-\dfrac{\partial A_c}{\partial t}$	$-\dfrac{\partial\phi_a}{\partial d}-\dfrac{\partial A_d}{\partial t}$			
$\left\{\begin{array}{l}\{3,17\}\\\{43,57\}\end{array}\right\}$	$-\dfrac{\partial\phi_a}{\partial b}-\dfrac{\partial A_b}{\partial t}$	$-\dfrac{\partial\phi_a}{\partial c}-\dfrac{\partial A_c}{\partial t}$	$-\dfrac{\partial\phi_a}{\partial d}-\dfrac{\partial A_d}{\partial t}$			
$\left\{\begin{array}{l}\{5,23\}\\45,63\end{array}\right\}$	$-\dfrac{\partial\phi_a}{\partial b}-\dfrac{\partial A_b}{\partial t}$				$-\dfrac{\partial\phi_a}{\partial g}-\dfrac{\partial A_g}{\partial t}$	$-\dfrac{\partial\phi_a}{\partial h}-\dfrac{\partial A_h}{\partial t}$
$\left\{\begin{array}{l}\{7,21\}\\\{47,61\}\end{array}\right\}$	$-\dfrac{\partial\phi_a}{\partial b}-\dfrac{\partial A_b}{\partial t}$				$-\dfrac{\partial\phi_a}{\partial g}-\dfrac{\partial A_g}{\partial t}$	$-\dfrac{\partial\phi_a}{\partial h}-\dfrac{\partial A_h}{\partial t}$
$\left\{\begin{array}{l}\{9,27\}\\\{33,51\}\end{array}\right\}$	$-\dfrac{\partial\phi_a}{\partial b}-\dfrac{\partial A_b}{\partial t}$	$-\dfrac{\partial\phi_a}{\partial c}-\dfrac{\partial A_c}{\partial t}$	$-\dfrac{\partial\phi_a}{\partial d}-\dfrac{\partial A_d}{\partial t}$			
$\left\{\begin{array}{l}\{11,25\}\\\{35,49\}\end{array}\right\}$	$-\dfrac{\partial\phi_a}{\partial b}-\dfrac{\partial A_b}{\partial t}$	$-\dfrac{\partial\phi_a}{\partial c}-\dfrac{\partial A_c}{\partial t}$	$-\dfrac{\partial\phi_a}{\partial d}-\dfrac{\partial A_d}{\partial t}$			
$\left\{\begin{array}{l}\{13,31\}\\\{37,55\}\end{array}\right\}$	$-\dfrac{\partial\phi_a}{\partial b}-\dfrac{\partial A_b}{\partial t}$				$-\dfrac{\partial\phi_a}{\partial g}-\dfrac{\partial A_g}{\partial t}$	$-\dfrac{\partial\phi_a}{\partial h}-\dfrac{\partial A_h}{\partial t}$
$\left\{\begin{array}{l}\{15,29\}\\\{39,53\}\end{array}\right\}$	$-\dfrac{\partial\phi_a}{\partial b}-\dfrac{\partial A_b}{\partial t}$				$-\dfrac{\partial\phi_a}{\partial g}-\dfrac{\partial A_g}{\partial t}$	$-\dfrac{\partial\phi_a}{\partial h}-\dfrac{\partial A_h}{\partial t}$
$\left\{\begin{array}{l}\{65,83\}\\\{105,123\}\end{array}\right\}$		$-\dfrac{\partial\phi_a}{\partial c}-\dfrac{\partial A_c}{\partial t}$		$-\dfrac{\partial\phi_a}{\partial f}-\dfrac{\partial A_f}{\partial t}$		$-\dfrac{\partial\phi_a}{\partial h}-\dfrac{\partial A_h}{\partial t}$
$\left\{\begin{array}{l}\{67,81\}\\\{107.121\}\end{array}\right\}$		$-\dfrac{\partial\phi_a}{\partial c}-\dfrac{\partial A_c}{\partial t}$		$-\dfrac{\partial\phi_a}{\partial f}-\dfrac{\partial A_f}{\partial t}$		$-\dfrac{\partial\phi_a}{\partial h}-\dfrac{\partial A_h}{\partial t}$
$\left\{\begin{array}{l}\{69,87\}\\\{109,127\}\end{array}\right\}$			$-\dfrac{\partial\phi_a}{\partial d}-\dfrac{\partial A_d}{\partial t}$	$-\dfrac{\partial\phi_a}{\partial f}-\dfrac{\partial A_f}{\partial t}$	$-\dfrac{\partial\phi_a}{\partial g}-\dfrac{\partial A_g}{\partial t}$	
$\left\{\begin{array}{l}\{71,85\}\\\{111,125\}\end{array}\right\}$			$-\dfrac{\partial\phi_a}{\partial d}-\dfrac{\partial A_d}{\partial t}$	$-\dfrac{\partial\phi_a}{\partial f}-\dfrac{\partial A_f}{\partial t}$	$-\dfrac{\partial\phi_a}{\partial g}-\dfrac{\partial A_g}{\partial t}$	
$\left\{\begin{array}{l}\{73,91\}\\\{97,115\}\end{array}\right\}$		$-\dfrac{\partial\phi_a}{\partial c}-\dfrac{\partial A_c}{\partial t}$		$-\dfrac{\partial\phi_a}{\partial f}-\dfrac{\partial A_f}{\partial t}$		$-\dfrac{\partial\phi_a}{\partial h}-\dfrac{\partial A_h}{\partial t}$

$\begin{Bmatrix} \{75,89\} \\ \{99,113\} \end{Bmatrix}$		$-\dfrac{\partial\phi_a}{\partial c}-\dfrac{\partial A_c}{\partial t}$		$-\dfrac{\partial\phi_a}{\partial f}-\dfrac{\partial A_f}{\partial t}$		$-\dfrac{\partial\phi_a}{\partial h}-\dfrac{\partial A_h}{\partial t}$
$\begin{Bmatrix} \{77,95\} \\ \{101,119\} \end{Bmatrix}$			$-\dfrac{\partial\phi_a}{\partial d}-\dfrac{\partial A_d}{\partial t}$	$-\dfrac{\partial\phi_a}{\partial f}-\dfrac{\partial A_f}{\partial t}$	$-\dfrac{\partial\phi_a}{\partial g}-\dfrac{\partial A_g}{\partial t}$	
$\begin{Bmatrix} \{79,93\} \\ \{103,117\} \end{Bmatrix}$			$-\dfrac{\partial\phi_a}{\partial d}-\dfrac{\partial A_d}{\partial t}$	$-\dfrac{\partial\phi_a}{\partial f}-\dfrac{\partial A_f}{\partial t}$	$-\dfrac{\partial\phi_a}{\partial g}-\dfrac{\partial A_g}{\partial t}$	

We note that, taken together, the sixty four copies of the $Cl_{3,0} \cong Cl_{1,2} = 1+3\sqrt{+1}+4\sqrt{-1}_{Non-Com}$ algebra have twice the number of E-fields as do either of the other two types of algebra.

Considering the three types of 8-dimensional algebra, taking the emergent E-fields of the form (25.2), we get:

$$Cl_{0,3} \sim 1+6\sqrt{-1}+\sqrt{+1}\dots\dots\dots\dots\dots\dots\dots 4 \ E-field$$

$$Cl_{2,1} \sim 1+5\sqrt{-1}+2\sqrt{+1}\dots\dots\dots\dots\dots\dots\dots 4 \ E-field \qquad (25.3)$$

$$Cl_{3,0} \sim Cl_{1,2} \sim 1+4\sqrt{-1}+3\sqrt{+1}\dots\dots\dots\dots\dots\dots 8 \ E-field$$

We see, (25.3), have three types of '8-dimensional electron'. Two of these types have the same amount of E-field and are indistinguishable in terms of total E-field. The other type has twice the total E-field of the other two types. This sounds like two down quarks and one up quark.

If the only components of these fields were $E_{[1,2]}$, $E_{[1,3]}$, $E_{[1,4]}$, we could claim an electron field, but there are also the components $E_{[1,6]}$, $E_{[1,7]}$, $E_{[1,8]}$. Are these another field? We do not know, but it might be a colour field.

Summary:

The sign of the $P_{5,1}$ parameter divides the 128 8-dimensional algebras into two sets. One of these two sets is 64 copies of a single type of algebra; the other of these sets is comprised of two different types of algebra. If we accept that we can observe only fields of the forms (25.1) & (25.2) in our 4-dimensional space-time, then we get two types of '8-dimensional electron' – let us call them quarks. One of these quarks is the electric particle of the $Cl_{3,0} \sim Cl_{1,2} \sim 1+4\sqrt{-1}+3\sqrt{+1}_{Non-Com}$ algebra and it has twice the electric field of the other two quarks which are the electric particles of the other two algebras. Electrically, the other two algebras are indistinguishable.

Anti-matter Quarks

The left-chiral quaternion matter potential which gave the electron field was:

$$\Phi_{L\chi}^{Matter} = \begin{bmatrix} \phi & -A_x & -A_y & -A_z \\ A_x & \phi & A_z & -A_y \\ A_y & -A_z & \phi & A_x \\ A_z & A_y & -A_x & \phi \end{bmatrix} \tag{26.1}$$

This, (26.1), gave the left-chiral electron field.

We also had the left-chiral quaternion anti-matter potential:

$$\Phi_{L\chi}^{Anti-Matter} = \begin{bmatrix} -\phi & A_x & A_y & A_z \\ -A_x & -\phi & -A_z & A_y \\ -A_y & A_z & -\phi & -A_x \\ -A_z & -A_y & A_x & -\phi \end{bmatrix} \tag{26.2}$$

This, (26.2), gave the left-chiral positron field.

We could form the anti-matter quaternion potential with negative real variable because we do not need to take the exponential to make the quaternion algebraic matrix form into a division algebra. Within the 8-dimensional algebras, we cannot take the real variable to be negative, and so there is no 8-dimensional anti-matter potential. We might therefore expect no 8-dimensional anti-matter. Things are not that simple.

Above, (24.12), we briefly mentioned emergent E-fields of the form:

$$\begin{bmatrix} Div & E_{[1,2]} & E_{[1,3]} & E_{[1,4]} & 0 & 0 & 0 & 0 \\ -E_{[1,2]} & Div & 0 & 0 & 0 & 0 & 0 & 0 \\ -E_{[1,3]} & 0 & Div & 0 & 0 & 0 & 0 & 0 \\ -E_{[1,4]} & 0 & 0 & Div & 0 & 0 & 0 & 0 \\ 0 & 0 & 0 & 0 & Div & -E_{[6,5]} & -E_{[7,5]} & -E_{[8,5]} \\ 0 & 0 & 0 & 0 & E_{[6,5]} & Div & 0 & 0 \\ 0 & 0 & 0 & 0 & E_{[7,5]} & 0 & Div & 0 \\ 0 & 0 & 0 & 0 & E_{[8,5]} & 0 & 0 & Div \end{bmatrix} \tag{26.3}$$

These fields, (26.3), seem to carry two 4-dimensional electric fields of opposite charge. We take this to be the electric charge part of the nature of matter and anti-matter in 8-dimensions; when we say '8-dimensions', we have in mind the quarks.

It is a mathematical fact that sometimes, cases 20 & 60, we see a perfect 'double copy' of a quaternion on the leading diagonal of a 8-dimensional algebra. Most of the cases of the 8-dimensional algebras do not have perfect copies of a 4-dimensional algebra on the leading diagonal. For the purposes of demonstrating this, we will set the $\{e, f, g, h\}$ variables to zero. We then have:

We have Case 2:

$$
\begin{bmatrix}
a & b & c & d & & & & \\
-b & a & d & -c & & & & \\
-c & -d & a & b & & & & \\
-d & c & -b & a & & & & \\
& & & & a & -b & -c & -d \\
& & & & b & a & d & -c \\
& & & & c & -d & a & b \\
& & & & d & c & -b & a
\end{bmatrix}
\equiv
\begin{bmatrix}
\begin{bmatrix} \mathbb{H}_{R\chi} \end{bmatrix} & \sim \\
\sim & \begin{bmatrix} \mathbb{H}_{L\chi} \end{bmatrix}^{*b*c*d}
\end{bmatrix}
\tag{26.4}
$$

Wherein we have used the notation $*b$ to indicate that the variable b is conjugated.

The other $Cl_{0,3} \sim 1 + 6\sqrt{-1} + \sqrt{+1}$ cases are:

2	$\begin{bmatrix} \begin{bmatrix} \mathbb{H}_{R\chi} \end{bmatrix} & \sim \\ \sim & \begin{bmatrix} \mathbb{H}_{L\chi} \end{bmatrix}^{*b*c*d} \end{bmatrix}$	4	$\begin{bmatrix} \begin{bmatrix} \mathbb{H}_{R\chi} \end{bmatrix} & \sim \\ \sim & \begin{bmatrix} \mathbb{H}_{L\chi} \end{bmatrix}^{*b} \end{bmatrix}$
10	$\begin{bmatrix} \begin{bmatrix} \mathbb{H}_{R\chi} \end{bmatrix} & \sim \\ \sim & \begin{bmatrix} \mathbb{H}_{R\chi} \end{bmatrix}^{*b*c} \end{bmatrix}$	12	$\begin{bmatrix} \begin{bmatrix} \mathbb{H}_{R\chi} \end{bmatrix} & \sim \\ \sim & \begin{bmatrix} \mathbb{H}_{R\chi} \end{bmatrix}^{*b*d} \end{bmatrix}$
18	$\begin{bmatrix} \begin{bmatrix} \mathbb{H}_{R\chi} \end{bmatrix} & \sim \\ \sim & \begin{bmatrix} \mathbb{H}_{R\chi} \end{bmatrix}^{*c*d} \end{bmatrix}$	20	$\begin{bmatrix} \begin{bmatrix} \mathbb{H}_{R\chi} \end{bmatrix} & \sim \\ \sim & \begin{bmatrix} \mathbb{H}_{R\chi} \end{bmatrix} \end{bmatrix}$ Perfect copy
26	$\begin{bmatrix} \begin{bmatrix} \mathbb{H}_{R\chi} \end{bmatrix} & \sim \\ \sim & \begin{bmatrix} \mathbb{H}_{L\chi} \end{bmatrix}^{*c} \end{bmatrix}$	28	$\begin{bmatrix} \begin{bmatrix} \mathbb{H}_{R\chi} \end{bmatrix} & \sim \\ \sim & \begin{bmatrix} \mathbb{H}_{L\chi} \end{bmatrix}^{*d} \end{bmatrix}$
34	$\begin{bmatrix} \begin{bmatrix} \mathbb{H}_{L\chi} \end{bmatrix} & \sim \\ \sim & \begin{bmatrix} \mathbb{H}_{L\chi} \end{bmatrix}^{*b*c} \end{bmatrix}$	36	$\begin{bmatrix} \begin{bmatrix} \mathbb{H}_{L\chi} \end{bmatrix} & \sim \\ \sim & \begin{bmatrix} \mathbb{H}_{L\chi} \end{bmatrix}^{*b*d} \end{bmatrix}$
42	$\begin{bmatrix} \begin{bmatrix} \mathbb{H}_{L\chi} \end{bmatrix} & \sim \\ \sim & \begin{bmatrix} \mathbb{H}_{R\chi} \end{bmatrix}^{*b*c*d} \end{bmatrix}$	44	$\begin{bmatrix} \begin{bmatrix} \mathbb{H}_{L\chi} \end{bmatrix} & \sim \\ \sim & \begin{bmatrix} \mathbb{H}_{R\chi} \end{bmatrix}^{*b} \end{bmatrix}$
50	$\begin{bmatrix} \begin{bmatrix} \mathbb{H}_{L\chi} \end{bmatrix} & \sim \\ \sim & \begin{bmatrix} \mathbb{H}_{R\chi} \end{bmatrix}^{*c} \end{bmatrix}$	52	$\begin{bmatrix} \begin{bmatrix} \mathbb{H}_{L\chi} \end{bmatrix} & \sim \\ \sim & \begin{bmatrix} \mathbb{H}_{R\chi} \end{bmatrix}^{*d} \end{bmatrix}$
58	$\begin{bmatrix} \begin{bmatrix} \mathbb{H}_{L\chi} \end{bmatrix} & \sim \\ \sim & \begin{bmatrix} \mathbb{H}_{L\chi} \end{bmatrix}^{*c*d} \end{bmatrix}$	60	$\begin{bmatrix} \begin{bmatrix} \mathbb{H}_{L\chi} \end{bmatrix} & \sim \\ \sim & \begin{bmatrix} \mathbb{H}_{L\chi} \end{bmatrix} \end{bmatrix}$ Perfect copy

If the electrical part of the nature of anti-matter in the quarks is due to the difference in the charges, directions, of electric fields on the leading diagonal, then we see that we do not have the simple matter/anti-matter relations of the electron and positron in the quark sector of particle physics. Indeed, the perfect copy cases 20 and 60 have no anti-matter at all.

There are similar results for the cases of the other algebras, but, most often, the A_3 algebras appear on the leading diagonals.

Of course, these are just individual algebras. It is the superimposition which we observe in our 4-dimensional space-time.

Summary:
We do not properly understand anti-matter in the quarks.

Chiral Quarks

We saw above that the case 60 algebra has two left-chiral quaternions on the leading diagonal; it would make sense to arbitrarily say that the case 60 algebra is left-chiral because this fits with the chirality of the quaternions. The case 60 algebraic matrix form is:

$$Case\ 60 = \begin{bmatrix} a & b & c & d & e & f & g & h \\ -b & a & -d & c & -f & e & -h & g \\ -c & d & a & -b & -g & h & e & -f \\ -d & -c & b & a & -h & -g & f & e \\ e & f & g & h & a & b & c & d \\ -f & e & -h & g & -b & a & -d & c \\ -g & h & e & -f & -c & d & a & -b \\ -h & -g & f & e & -d & -c & b & a \end{bmatrix} \sim \begin{bmatrix} \mathbb{H}_{L\chi} & \mathbb{H}_{L\chi} \\ \mathbb{H}_{L\chi} & \mathbb{H}_{L\chi} \end{bmatrix} \qquad (27.1)$$

Similarly, the case 20 algebra is wholly right-chiral:

$$Case\ 20 = \begin{bmatrix} a & b & c & d & e & f & g & h \\ -b & a & d & -c & -f & e & h & -g \\ -c & -d & a & b & -g & -h & e & f \\ -d & c & -b & a & -h & g & -f & e \\ e & f & g & h & a & b & c & d \\ -f & e & h & -g & -b & a & d & -c \\ -g & -h & e & f & -c & -d & a & b \\ -h & g & -f & e & -d & c & -b & a \end{bmatrix} \sim \begin{bmatrix} \mathbb{H}_{R\chi} & \mathbb{H}_{R\chi} \\ \mathbb{H}_{R\chi} & \mathbb{H}_{R\chi} \end{bmatrix} \qquad (27.2)$$

Surely, we now have a definition of left-chirality and of right-chirality within the 8-dimensional $C_2 \times C_2 \times C_2$ algebras.

The case 2 algebra has the same commutation relations as the case 20 algebra, and so surely the case 2 algebra is right-chiral. Similarly, since the case 42 algebra has the same commutation relations as the case 60 algebra, surely the case 42 algebra is left-chiral. We have:

$$Case\ 2 = \begin{bmatrix} a & b & c & d & e & f & g & h \\ -b & a & d & -c & -f & e & h & -g \\ -c & -d & a & b & -g & -h & e & f \\ -d & c & -b & a & -h & g & -f & e \\ e & f & g & h & a & b & c & d \\ -f & e & h & -g & -b & a & -d & c \\ -g & -h & e & f & -c & d & a & -b \\ -h & g & -f & e & -d & -c & b & a \end{bmatrix} \sim \begin{bmatrix} \mathbb{H}_{R\chi} & M\mathbb{H}_{L\chi} \\ M\mathbb{H}_{R\chi} & \mathbb{H}_{L\chi} \end{bmatrix} : M = \begin{bmatrix} 1 & 0 & 0 & 0 \\ 0 & -1 & 0 & 0 \\ 0 & 0 & -1 & 0 \\ 0 & 0 & 0 & -1 \end{bmatrix}$$

$$(27.3)$$

Clearly, even within an algebra marriage, the quaternion type chirality is violated. This is violation of CP invariance. The commutation relations are not sufficient to determine chirality in the 8-dimensional algebras.

We also have:

$$Case\ 42 = \begin{bmatrix} a & b & c & d & e & f & g & h \\ -b & a & -d & c & f & -e & -h & g \\ -c & d & a & -b & g & h & -e & -f \\ -d & -c & b & a & h & -g & f & -e \\ e & -f & -g & -h & a & -b & -c & -d \\ -f & -e & -h & g & b & a & -d & c \\ -g & h & -e & -f & c & d & a & -b \\ -h & -g & f & -e & d & -c & b & a \end{bmatrix} \sim \begin{bmatrix} \mathbb{H}_{L\chi} & M\mathbb{H}_{R\chi} \\ M\mathbb{H}_{L\chi} & \mathbb{H}_{R\chi} \end{bmatrix} : M = \begin{bmatrix} 1 & 0 & 0 & 0 \\ 0 & -1 & 0 & 0 \\ 0 & 0 & -1 & 0 \\ 0 & 0 & 0 & -1 \end{bmatrix}$$

$$(27.4)$$

Chapter 28

The B-fields of the 8-dimensional Algebras

Now let us consider the B-fields of the 8-dimensional algebras.

The $Cl_{0,3}$ B-fields:

The B-fields of the $Cl_{0,3} \sim 1 + 6\sqrt{-1} + \sqrt{+1}$ algebras are given below. The variables in the differentials are the same in each algebra; only the signs differ, and so we show only the signs.

In every case, we have $B_{[1,1]} = 0$ & $B_{[1,5]} = 0$

Marriage	Case	$B_{[1,2]}$	$B_{[1,3]}$	$B_{[1,4]}$
	2	$\dfrac{\partial A_c}{\partial d} - \dfrac{\partial A_d}{\partial c} + \dfrac{\partial A_g}{\partial h} - \dfrac{\partial A_h}{\partial g}$	$-\dfrac{\partial A_b}{\partial d} + \dfrac{\partial A_d}{\partial b} - \dfrac{\partial A_f}{\partial h} + \dfrac{\partial A_h}{\partial f}$	$\dfrac{\partial A_b}{\partial c} - \dfrac{\partial A_c}{\partial b} + \dfrac{\partial A_f}{\partial g} - \dfrac{\partial A_g}{\partial f}$
$\begin{Bmatrix} \{2,20\} \\ \{42,60\} \end{Bmatrix}$		$(+,-,+,-)$	$(-,+,-,+)$	$(+,-,+,-)$
	20	$(+,-,+,-)$	$(-,+,-,+)$	$(+,-,+,-)$
	42	$(-,+,-,+)$	$(+,-,+,-)$	$(-,+,-,+)$
	60	$(-,+,-,+)$	$(+,-,+,-)$	$(-,+,-,+)$
	4	$(+,-,+,-)$	$(-,+,+,-)$	$(+,-,-,+)$
$\begin{Bmatrix} \{4,18\} \\ \{44,58\} \end{Bmatrix}$	18	$(+,-,+,-)$	$(-,+,+,-)$	$(+,-,-,+)$
	44	$(-,+,-,+)$	$(+,-,-,+)$	$(-,+,+,-)$
	58	$(-,+,-,+)$	$(+,-,-,+)$	$(-,+,+,-)$
	10	$(+,-,-,+)$	$(-,+,+,-)$	$(+,-,+,-)$
$\begin{Bmatrix} \{10,28\} \\ \{34,52\} \end{Bmatrix}$	28	$(+,-,-,+)$	$(-,+,+,-)$	$(+,-,+,-)$
	34	$(-,+,+,-)$	$(+,-,-,+)$	$(-,+,-,+)$
	52	$(-,+,+,-)$	$(+,-,-,+)$	$(-,+,-,+)$
	12	$(+,-,+,-)$	$(-,+,+,-)$	$(+,-,+,-)$
$\begin{Bmatrix} \{12,26\} \\ \{36,50\} \end{Bmatrix}$	26	$(+,-,+,-)$	$(-,+,+,-)$	$(+,-,+,-)$
	36	$(-,+,-,+)$	$(+,-,-,+)$	$(-,+,-,+)$
	50	$(-,+,-,+)$	$(+,-,-,+)$	$(-,+,-,+)$

B-Fields of the $Cl_{0,3} \sim 1 + 6\sqrt{-1} + \sqrt{+1}$ algebras – page 1

Marriage	Case	$B_{[1,6]}$	$B_{[1,7]}$	$B_{[1,8]}$
		B-Fields of the $Cl_{0,3} \sim 1+6\sqrt{-1}+\sqrt{+1}$ algebras – page 2		
$\left\{\begin{matrix}\{2,20\}\\\{42,60\}\end{matrix}\right\}$	2	$\dfrac{\partial A_c}{\partial h}-\dfrac{\partial A_d}{\partial g}+\dfrac{\partial A_g}{\partial d}-\dfrac{\partial A_h}{\partial c}$ $(+,-,+,-)$	$-\dfrac{\partial A_b}{\partial h}+\dfrac{\partial A_d}{\partial f}-\dfrac{\partial A_f}{\partial d}+\dfrac{\partial A_h}{\partial b}$ $(-,+,-,+)$	$\dfrac{\partial A_b}{\partial g}-\dfrac{\partial A_c}{\partial f}+\dfrac{\partial A_f}{\partial c}-\dfrac{\partial A_g}{\partial b}$ $(+,-,+,-)$
	20	$(+,-,+,-)$	$(-,+,-,+)$	$(+,-,+,-)$
	42	$(-,+,-,+)$	$(+,-,+,-)$	$(-,+,-,+)$
	60	$(-,+,-,+)$	$(+,-,+,-)$	$(-,+,-,+)$
$\left\{\begin{matrix}\{4,18\}\\\{44,58\}\end{matrix}\right\}$	4	$(-,+,-,+)$	$(-,-,+,+)$	$(+,+,-,-)$
	18	$(-,+,-,+)$	$(-,-,+,+)$	$(+,+,-,-)$
	44	$(+,-,+,-)$	$(+,+,-,-)$	$(-,-,+,+)$
	58	$(+,-,+,-)$	$(+,+,-,-)$	$(-,-,+,+)$
$\left\{\begin{matrix}\{10,28\}\\\{34,52\}\end{matrix}\right\}$	10	$(-,-,+,+)$	$(+,+,-,-)$	$(-,+,-,+)$
	28	$(-,-,+,+)$	$(+,+,-,-)$	$(-,+,-,+)$
	34	$(+,+,-,-)$	$(-,-,+,+)$	$(+,-,+,-)$
	52	$(+,+,-,-)$	$(-,-,+,+)$	$(+,-,+,-)$
$\left\{\begin{matrix}\{12,26\}\\\{36,50\}\end{matrix}\right\}$	12	$(+,+,-,-)$	$(+,-,+,-)$	$(-,-,+,+)$
	26	$(+,+,-,-)$	$(+,-,+,-)$	$(-,-,+,+)$
	36	$(-,-,+,+)$	$(-,+,-,+)$	$(+,+,-,-)$
	50	$(-,-,+,+)$	$(-,+,-,+)$	$(+,+,-,-)$

The above tables are a little deceptive in that they take no account of the distribution of minus signs within the different matrices.

The emergent B-fields:

We recall the quaternion B-field:

$$B_t = 0$$
$$B_x = \frac{\partial A_z}{\partial y} - \frac{\partial A_y}{\partial z}$$
$$B_y = \frac{\partial A_x}{\partial z} - \frac{\partial A_z}{\partial x} \tag{28.1}$$
$$B_z = \frac{\partial A_y}{\partial x} - \frac{\partial A_x}{\partial y}$$

And:

$$B^{Matter}_{\mathbb{H}_{L\chi}} = \begin{bmatrix} 0 & B_{[1,2]} & B_{[1,3]} & B_{[1,4]} \\ -B_{[1,2]} & 0 & -B_{[1,4]} & B_{[1,3]} \\ -B_{[1,3]} & B_{[1,4]} & 0 & -B_{[1,2]} \\ -B_{[1,4]} & -B_{[1,3]} & B_{[1,2]} & 0 \end{bmatrix} \tag{28.2}$$

If we add all sixteen B-fields of the $Cl_{0,3} \sim 1 + 6\sqrt{-1} + \sqrt{+1}$ algebras, we get:

$$B_{[2,3]} = 16\left(\frac{\partial A_b}{\partial c} - \frac{\partial A_c}{\partial b}\right) \qquad B_{[6,7]} = 16\left(\frac{\partial A_f}{\partial g} - \frac{\partial A_g}{\partial f}\right)$$

$$B_{[2,4]} = 16\left(\frac{\partial A_b}{\partial d} - \frac{\partial A_d}{\partial b}\right) \qquad B_{[6,8]} = 16\left(\frac{\partial A_f}{\partial h} - \frac{\partial A_h}{\partial f}\right) \tag{28.3}$$

$$B_{[3,4]} = 16\left(\frac{\partial A_c}{\partial d} - \frac{\partial A_d}{\partial c}\right) \qquad B_{[7,8]} = 16\left(\frac{\partial A_g}{\partial h} - \frac{\partial A_h}{\partial g}\right)$$

And:

$$B_{[2,7]} = 16\left(\frac{\partial A_b}{\partial g} - \frac{\partial A_g}{\partial b}\right) \quad B_{[3,6]} = 16\left(\frac{\partial A_c}{\partial f} - \frac{\partial A_f}{\partial c}\right)$$

$$B_{[2,8]} = 16\left(\frac{\partial A_b}{\partial h} - \frac{\partial A_h}{\partial b}\right) \quad B_{[4,6]} = 16\left(\frac{\partial A_d}{\partial f} - \frac{\partial A_f}{\partial d}\right) \tag{28.4}$$

$$B_{[3,8]} = 16\left(\frac{\partial A_c}{\partial h} - \frac{\partial A_h}{\partial c}\right) \quad B_{[4,7]} = 16\left(\frac{\partial A_d}{\partial g} - \frac{\partial A_g}{\partial d}\right)$$

And:

$$B^{Emergent}_{Cl_{0,3} \sim 1+6\sqrt{-1}+\sqrt{+1}} = \begin{bmatrix} 0 & 0 & 0 & 0 & 0 & 0 & 0 & 0 \\ 0 & 0 & B_{[2,3]} & B_{[2,4]} & 0 & 0 & B_{[2,7]} & B_{[2,8]} \\ 0 & -B_{[2,3]} & 0 & B_{[3,4]} & 0 & B_{[3,6]} & 0 & B_{[3,8]} \\ 0 & -B_{[2,4]} & -B_{[3,4]} & 0 & 0 & B_{[4,6]} & B_{[4,7]} & 0 \\ 0 & 0 & 0 & 0 & 0 & 0 & 0 & 0 \\ 0 & 0 & -B_{[3,6]} & -B_{[4,6]} & 0 & 0 & B_{[6,7]} & B_{[6,8]} \\ 0 & -B_{[2,7]} & 0 & -B_{[4,7]} & 0 & -B_{[6,7]} & 0 & B_{[7,8]} \\ 0 & -B_{[2,8]} & -B_{[3,8]} & 0 & 0 & -B_{[6,8]} & -B_{[7,8]} & 0 \end{bmatrix} \tag{28.5}$$

Looking at the top left-hand 4×4 corner, we see that the distribution of minus signs does not seem to have a definite chirality:

$$B^{Emergent}_{Cl_{0,3}\sim1+6\sqrt{-1}+\sqrt{+1}} \sim \begin{bmatrix} 0 & 0 & 0 & 0 & 0 & 0 & 0 & 0 \\ 0 & 0 & B_{[2,3]} & B_{[2,4]} & 0 & 0 & \sim & \sim \\ 0 & -B_{[2,3]} & 0 & B_{[3,4]} & 0 & \sim & 0 & \sim \\ 0 & -B_{[2,4]} & -B_{[3,4]} & 0 & 0 & \sim & \sim & 0 \\ 0 & 0 & 0 & 0 & 0 & 0 & 0 & 0 \\ 0 & 0 & \sim & \sim & 0 & 0 & \sim & \sim \\ 0 & \sim & 0 & \sim & 0 & \sim & 0 & \sim \\ 0 & \sim & \sim & 0 & 0 & \sim & \sim & 0 \end{bmatrix} \quad (28.6)$$

There is no 4-dimensional non-commutative $C_2 \times C_2$ algebra with this distribution of minus signs.

Comparing (28.4) with the quaternion B-field, we see that we can put:

$$B_{[2,3]} = \frac{\partial A_x}{\partial y} - \frac{\partial A_y}{\partial x} = -B_z$$

$$B_{[2,4]} = \frac{\partial A_x}{\partial z} - \frac{\partial A_z}{\partial x} = B_y \quad (28.7)$$

$$B_{[3,4]} = \frac{\partial A_y}{\partial z} - \frac{\partial A_z}{\partial y} = -B_x$$

Substituting these, (28.7), into the emergent B-field, (28.5), gives:

$$B^{Emergent}_{Cl_{0,3}\sim1+6\sqrt{-1}+\sqrt{+1}} \sim \begin{bmatrix} 0 & 0 & 0 & 0 & 0 & 0 & 0 & 0 \\ 0 & 0 & -B_z & B_y & 0 & 0 & \sim & \sim \\ 0 & B_z & 0 & -B_x & 0 & \sim & 0 & \sim \\ 0 & -B_y & B_x & 0 & 0 & \sim & \sim & 0 \\ 0 & 0 & 0 & 0 & 0 & 0 & 0 & 0 \\ 0 & 0 & \sim & \sim & 0 & 0 & \sim & \sim \\ 0 & \sim & 0 & \sim & 0 & \sim & 0 & \sim \\ 0 & \sim & \sim & 0 & 0 & \sim & \sim & 0 \end{bmatrix} \quad (28.8)$$

We see that the top left-hand 4×4 corner is of the form of a left-chiral quaternion.

To make the bottom right-hand 4×4 corner of (28.5) into a left-chiral quaternion, we need to put:

$$B_{[6,7]} = \frac{\partial A_f}{\partial g} - \frac{\partial A_g}{\partial f} = -B_z$$

$$B_{[6,8]} = \frac{\partial A_f}{\partial h} - \frac{\partial A_h}{\partial f} = B_y \quad (28.9)$$

$$B_{[7,8]} = \frac{\partial A_g}{\partial h} - \frac{\partial A_h}{\partial g} = -B_x$$

But, if we reverse the signs in (28.9) to give:

$$B_{[6,7]} = \frac{\partial A_f}{\partial g} - \frac{\partial A_g}{\partial f} = B_z$$

$$B_{[6,8]} = \frac{\partial A_f}{\partial h} - \frac{\partial A_h}{\partial f} = -B_y \qquad (28.10)$$

$$B_{[7,8]} = \frac{\partial A_g}{\partial h} - \frac{\partial A_h}{\partial g} = B_x$$

We get a right-chiral quaternion in the bottom right-hand 4×4 corner of (28.5). The chirality of the emergent B-field is not properly defined in terms of quaternion chirality.

Opinion:

We do not properly understand the chirality of the emergent B-field of the 8-dimensional $C_2 \times C_2 \times C_2$ algebras. We opine that, analogously to the 4-dimensional case, the emergent B-field of the 8-dimensional $C_2 \times C_2 \times C_2$ algebras is a neutrino type of field, but we cannot say that this neutrino field is always left-handed because we have not been able to define chirality unambiguously within the $C_2 \times C_2 \times C_2$ algebras. We are predicting a kind of 'quark neutrino' field. Could this be the gluons? We do not know.

Summary of the Chirality of the 8-dimensional Algebras

We do not understand the nature of the 8-dimensional chirality.

There are algebras like the case 60 algebra, (27.1), which are clearly left-chiral as judged by comparison with the chirality of the quaternions; this algebra has an opposite, the case 20 algebra, (27.2), which is clearly right-chiral as judged by comparison with the chirality of the quaternions. We must remember that each type of algebra has its own chirality; the chirality of the A_3 algebras is spatial parity; the chirality of the quaternions takes two forms as is evident in the E-fields and the B-fields; these two forms are charge conjugation for the E-fields and CP invariance for the B-field. We seem to have a much more complicated form of chirality for each of the three 8-dimensional algebras.

Commutation relations alone seem to be insufficient to determine chirality in the 8-dimensional algebras. The case 2 algebra, (27.3), has exactly the same commutation relations as the case 20 algebra, but we cannot say that, as judged by the chirality of the quaternions, that the case 2 algebra is a left-chiral algebra.

We certainly have pairs of algebras with opposite commutation relations, like the case 2 algebra and the case 42 algebra, (27.4), or the case 20 algebra and the case 60 algebra, but these pairs are paired together to form an algebra marriage. It seems that, by quaternion standards, we have two types of left-chirality matched by two types of right-chirality. By commutation relations alone, we could pair together the case 2 algebra and the case 60 algebra rather than the case 20 algebra and the case 60 algebra, yet, looking at the algebraic matrix forms of these algebras, aesthetically, we prefer the case 20 and case 60 pairing.

We clearly have violation of CP invariance within the 8-dimensional algebras. We are aware that violation of CP invariance has been observed within the quark sector of the Standard Model of particle physics. We do not have violation of CP invariance within the 4-dimensional algebras. We are aware that violation of CP invariance has not been observed in the electron neutrino sector of the Standard Model of particle physics; it has been sought.

We must leave the reader without a clear understanding of chirality in the 8-dimensional algebras.

Almost done:

We have now finished, but not yet completed, our exploration of the 8-dimensional $C_2 \times C_2 \times C_2$ algebras, the 8-dimensional Clifford algebras. We now have only one more step to take. In the next chapter, we will show that the 16-dimensional $C_2 \times C_2 \times C_2 \times C_2$ algebras, the 16-dimensional Clifford algebras, are not chiral algebras.

Chapter 30

The 16-dimensional Algebras

As originally formulated by Clifford, the 16-dimensional Clifford algebras are chiral algebras with relations based upon:

$$\vec{e_a}\vec{e_b} = -\vec{e_b}\vec{e_a} \tag{30.1}$$

However, when we present the 16-dimensional Clifford algebras as division algebras derived from the $C_2 \times C_2 \times C_2 \times C_2$ group, remarkably, this chirality disappears.

Commutators and anti-commutators again:

Earlier, we defined commutation relations in terms of the commutator and the anti-commutator:

$$Commutator = [B \quad C] = BC - CB \qquad\qquad Anti-commutator = \{B \quad C\} = BC + CB \tag{30.2}$$

We defined a chiral algebra to be one in which, for any two variables in that algebra, we have either a non-zero commutator and a zero anti-commutator or a non-zero anti-commutator and a zero commutator. We required at least one non-zero commutator. An example is the left-chiral quaternion algebra:

$$\mathbb{H}_{L\chi} = \begin{bmatrix} a & b & c & d \\ -b & a & -d & c \\ -c & d & a & -b \\ -d & -c & b & a \end{bmatrix} \qquad [b \quad c] = d - -d = 2d \qquad \{b \quad c\} = d + -d = 0 \tag{30.3}$$

Other examples are any of the 8-dimensional $C_2 \times C_2 \times C_2$ algebras.

Within a chiral algebra, we have no pairs of variables which have a non-zero commutator and a non-zero anti-commutator.

Now consider the j variable of a 16-dimensional $C_2 \times C_2 \times C_2 \times C_2$ algebra:

$$\begin{bmatrix}
0 & 0 & 0 & 0 & 0 & 0 & 0 & 0 & 0 & j & 0 & 0 & 0 & 0 & 0 & 0 \\
0 & 0 & 0 & 0 & 0 & 0 & 0 & 0 & j & 0 & 0 & 0 & 0 & 0 & 0 & 0 \\
0 & 0 & 0 & 0 & 0 & 0 & 0 & 0 & 0 & 0 & 0 & j & 0 & 0 & 0 & 0 \\
0 & 0 & 0 & 0 & 0 & 0 & 0 & 0 & 0 & 0 & j & 0 & 0 & 0 & 0 & 0 \\
0 & 0 & 0 & 0 & 0 & 0 & 0 & 0 & 0 & 0 & 0 & 0 & 0 & j & 0 & 0 \\
0 & 0 & 0 & 0 & 0 & 0 & 0 & 0 & 0 & 0 & 0 & 0 & j & 0 & 0 & 0 \\
0 & 0 & 0 & 0 & 0 & 0 & 0 & 0 & 0 & 0 & 0 & 0 & 0 & 0 & 0 & j \\
0 & 0 & 0 & 0 & 0 & 0 & 0 & 0 & 0 & 0 & 0 & 0 & 0 & 0 & -j & 0 \\
0 & -j & 0 & 0 & 0 & 0 & 0 & 0 & 0 & 0 & 0 & 0 & 0 & 0 & 0 & 0 \\
-j & 0 & 0 & 0 & 0 & 0 & 0 & 0 & 0 & 0 & 0 & 0 & 0 & 0 & 0 & 0 \\
0 & 0 & 0 & -j & 0 & 0 & 0 & 0 & 0 & 0 & 0 & 0 & 0 & 0 & 0 & 0 \\
0 & 0 & -j & 0 & 0 & 0 & 0 & 0 & 0 & 0 & 0 & 0 & 0 & 0 & 0 & 0 \\
0 & 0 & 0 & 0 & 0 & -j & 0 & 0 & 0 & 0 & 0 & 0 & 0 & 0 & 0 & 0 \\
0 & 0 & 0 & 0 & -j & 0 & 0 & 0 & 0 & 0 & 0 & 0 & 0 & 0 & 0 & 0 \\
0 & 0 & 0 & 0 & 0 & 0 & 0 & j & 0 & 0 & 0 & 0 & 0 & 0 & 0 & 0 \\
0 & 0 & 0 & 0 & 0 & 0 & -j & 0 & 0 & 0 & 0 & 0 & 0 & 0 & 0 & 0
\end{bmatrix} \tag{30.4}$$

And the n variable of the same algebra;

$$\begin{bmatrix}
0 & 0 & 0 & 0 & 0 & 0 & 0 & 0 & 0 & 0 & 0 & 0 & 0 & n & 0 & 0 \\
0 & 0 & 0 & 0 & 0 & 0 & 0 & 0 & 0 & 0 & 0 & 0 & n & 0 & 0 & 0 \\
0 & 0 & 0 & 0 & 0 & 0 & 0 & 0 & 0 & 0 & 0 & 0 & 0 & 0 & 0 & -n \\
0 & 0 & 0 & 0 & 0 & 0 & 0 & 0 & 0 & 0 & 0 & 0 & 0 & 0 & -n & 0 \\
0 & 0 & 0 & 0 & 0 & 0 & 0 & 0 & 0 & n & 0 & 0 & 0 & 0 & 0 & 0 \\
0 & 0 & 0 & 0 & 0 & 0 & 0 & 0 & n & 0 & 0 & 0 & 0 & 0 & 0 & 0 \\
0 & 0 & 0 & 0 & 0 & 0 & 0 & 0 & 0 & 0 & 0 & -n & 0 & 0 & 0 & 0 \\
0 & 0 & 0 & 0 & 0 & 0 & 0 & 0 & 0 & 0 & -n & 0 & 0 & 0 & 0 & 0 \\
0 & 0 & 0 & 0 & 0 & -n & 0 & 0 & 0 & 0 & 0 & 0 & 0 & 0 & 0 & 0 \\
0 & 0 & 0 & 0 & -n & 0 & 0 & 0 & 0 & 0 & 0 & 0 & 0 & 0 & 0 & 0 \\
0 & 0 & 0 & 0 & 0 & 0 & 0 & n & 0 & 0 & 0 & 0 & 0 & 0 & 0 & 0 \\
0 & 0 & 0 & 0 & 0 & 0 & n & 0 & 0 & 0 & 0 & 0 & 0 & 0 & 0 & 0 \\
0 & -n & 0 & 0 & 0 & 0 & 0 & 0 & 0 & 0 & 0 & 0 & 0 & 0 & 0 & 0 \\
-n & 0 & 0 & 0 & 0 & 0 & 0 & 0 & 0 & 0 & 0 & 0 & 0 & 0 & 0 & 0 \\
0 & 0 & 0 & n & 0 & 0 & 0 & 0 & 0 & 0 & 0 & 0 & 0 & 0 & 0 & 0 \\
0 & 0 & n & 0 & 0 & 0 & 0 & 0 & 0 & 0 & 0 & 0 & 0 & 0 & 0 & 0
\end{bmatrix} \tag{30.5}$$

The commutator is non-zero (there are just four copies of the e variable):

$$[j \quad n] = \begin{bmatrix}
0 & 0 & 0 & 0 & 0 & 0 & 0 & 0 & 0 & 0 & 0 & 0 & 0 & 0 & 0 & 0 \\
0 & 0 & 0 & 0 & 0 & 0 & 0 & 0 & 0 & 0 & 0 & 0 & 0 & 0 & 0 & 0 \\
0 & 0 & 0 & 0 & 0 & 0 & 0 & 0 & 0 & 0 & 0 & 0 & 0 & 0 & 0 & 0 \\
0 & 0 & 0 & 0 & 0 & 0 & 0 & e & 0 & 0 & 0 & 0 & 0 & 0 & 0 & 0 \\
0 & 0 & 0 & 0 & 0 & 0 & 0 & 0 & 0 & 0 & 0 & 0 & 0 & 0 & 0 & 0 \\
0 & 0 & 0 & 0 & 0 & 0 & 0 & 0 & 0 & 0 & 0 & 0 & 0 & 0 & 0 & 0 \\
0 & 0 & 0 & 0 & 0 & 0 & 0 & 0 & 0 & 0 & 0 & 0 & 0 & 0 & 0 & 0 \\
0 & 0 & 0 & -e & 0 & 0 & 0 & 0 & 0 & 0 & 0 & 0 & 0 & 0 & 0 & 0 \\
0 & 0 & 0 & 0 & 0 & 0 & 0 & 0 & 0 & 0 & 0 & 0 & 0 & 0 & 0 & 0 \\
0 & 0 & 0 & 0 & 0 & 0 & 0 & 0 & 0 & 0 & 0 & 0 & 0 & 0 & 0 & 0 \\
0 & 0 & 0 & 0 & 0 & 0 & 0 & 0 & 0 & 0 & 0 & 0 & 0 & 0 & e & 0 \\
0 & 0 & 0 & 0 & 0 & 0 & 0 & 0 & 0 & 0 & 0 & 0 & 0 & 0 & 0 & 0 \\
0 & 0 & 0 & 0 & 0 & 0 & 0 & 0 & 0 & 0 & 0 & 0 & 0 & 0 & 0 & 0 \\
0 & 0 & 0 & 0 & 0 & 0 & 0 & 0 & 0 & 0 & 0 & 0 & 0 & 0 & 0 & 0 \\
0 & 0 & 0 & 0 & 0 & 0 & 0 & 0 & 0 & 0 & -e & 0 & 0 & 0 & 0 & 0 \\
0 & 0 & 0 & 0 & 0 & 0 & 0 & 0 & 0 & 0 & 0 & 0 & 0 & 0 & 0 & 0
\end{bmatrix} \quad (30.6)$$

The anti-commutator is also non-zero:

$$\{j \quad n\} = \begin{bmatrix}
0 & 0 & 0 & 0 & -e & 0 & 0 & 0 & 0 & 0 & 0 & 0 & 0 & 0 & 0 & 0 \\
0 & 0 & 0 & 0 & 0 & -e & 0 & 0 & 0 & 0 & 0 & 0 & 0 & 0 & 0 & 0 \\
0 & 0 & 0 & 0 & 0 & 0 & e & 0 & 0 & 0 & 0 & 0 & 0 & 0 & 0 & 0 \\
0 & 0 & 0 & 0 & 0 & 0 & 0 & 0 & 0 & 0 & 0 & 0 & 0 & 0 & 0 & 0 \\
-e & 0 & 0 & 0 & 0 & 0 & 0 & 0 & 0 & 0 & 0 & 0 & 0 & 0 & 0 & 0 \\
0 & -e & 0 & 0 & 0 & 0 & 0 & 0 & 0 & 0 & 0 & 0 & 0 & 0 & 0 & 0 \\
0 & 0 & e & 0 & 0 & 0 & 0 & 0 & 0 & 0 & 0 & 0 & 0 & 0 & 0 & 0 \\
0 & 0 & 0 & 0 & 0 & 0 & 0 & 0 & 0 & 0 & 0 & 0 & 0 & 0 & 0 & 0 \\
0 & 0 & 0 & 0 & 0 & 0 & 0 & 0 & 0 & 0 & 0 & 0 & -e & 0 & 0 & 0 \\
0 & 0 & 0 & 0 & 0 & 0 & 0 & 0 & 0 & 0 & 0 & 0 & 0 & -e & 0 & 0 \\
0 & 0 & 0 & 0 & 0 & 0 & 0 & 0 & 0 & 0 & 0 & 0 & 0 & 0 & 0 & 0 \\
0 & 0 & 0 & 0 & 0 & 0 & 0 & 0 & 0 & 0 & 0 & 0 & 0 & 0 & 0 & e \\
0 & 0 & 0 & 0 & 0 & 0 & 0 & 0 & -e & 0 & 0 & 0 & 0 & 0 & 0 & 0 \\
0 & 0 & 0 & 0 & 0 & 0 & 0 & 0 & 0 & -e & 0 & 0 & 0 & 0 & 0 & 0 \\
0 & 0 & 0 & 0 & 0 & 0 & 0 & 0 & 0 & 0 & 0 & 0 & 0 & 0 & 0 & 0 \\
0 & 0 & 0 & 0 & 0 & 0 & 0 & 0 & 0 & 0 & 0 & e & 0 & 0 & 0 & 0
\end{bmatrix} \quad (30.7)$$

Sorry for the large matrices, but the point is clear. This 16-dimensional algebra is not a chiral algebra because it has at least two variables whose commutator and whose anti-commutator are both non-zero.

Since all $C_2 \times C_2 \times C_2 \times C_2 \times ...$ algebras of dimension greater than sixteen include 16-dimensional algebras as sub-algebras, then all $C_2 \times C_2 \times C_2 \times C_2 \times ...$ algebras of dimension greater than sixteen are not chiral algebras. We have run out of chirality.

Some sets of variables are mutually chiral:

Although the full algebra is not chiral, there are four sets of four variables which are mutually chiral with each other member of the set but which are not chiral with any variable outside of the set. These four sets of mutually chiral variables are:

$$\{a,b,c,d\}, \quad \{e,f,g,h\}, \quad \{i,j,k,l\}, \quad \{m,n,o,p\} \tag{30.8}$$

Of course, only the set containing the real variable, a, is a sub-algebra, but it is remarkable that the $\{e,i,m\}$ variables 'think they are' a real variable.

We have all chiral algebras:

All division algebras derive from finite groups. Only non-commutative division algebras that derive from commutative groups may be chiral algebras. The only such groups are the $C_2 \times C_2 \times ...$ groups and groups like $C_2 \times C_6$ which have these types of groups as sub-groups. Of these algebras, only the 4-dimensional non-commutative $C_2 \times C_2$ algebras and the 8-dimensional $C_2 \times C_2 \times C_2$ algebras can be manifest in our 4-dimensional space-time.

The conventional Standard Model:

The Standard Model of particle physics is built on the unitary Lie groups $U(1)$, $SU(2)$, and $SU(3)$. Why stop at $SU(3)$? Why not continue with $SU(4)$ and $SU(5)$? No-one can give an answer. If we replace the unitary Lie groups with the Clifford algebras in their division algebra forms as derived from the $C_2 \times C_2 \times ...$ groups, then we run out of chiral algebras at exactly the right point.

Summary:

The 2-dimensional C_2 algebras are not chiral because they are both commutative.

The 4-dimensional non-commutative $C_2 \times C_2$ algebras are chiral.

The 8-dimensional non-commutative $C_2 \times C_2 \times C_2$ algebras are chiral.

The 16-dimensional non-commutative $C_2 \times C_2 \times C_2 \times C_2$ algebras are not chiral.

The only chiral algebras in the whole universe which can be manifest in our 4-dimensional space-time are the 4-dimensional non-commutative $C_2 \times C_2$ algebras and the 8-dimensional non-commutative $C_2 \times C_2 \times C_2$ algebras.

This is a remarkable result.

The weird 16-dimensional variables:

It is suprising that most pairs of variables in the 16-dimensional algebras are 'half commutative' and half non-commutative'. Looking at (30.6) & (30.7), we might say that the $\{j, n\}$ variables are 75% commutative and 25% non-commutative. Of course, we need only the tiniest tad of non-commutativity to be conventionally not commutative.

The C₂ x C₆ Group

The $C_2 \times C_6$ group holds chiral algebras.

The $C_2 \times C_6$ group is a commutative group with three C_2 subgroups, one C_3 subgroup, one $C_2 \times C_2$ subgroup, and three C_6 subgroups.

The Standard Form Cayley table of the $C_2 \times C_6$ group is:

$$C_2 \times C_6 \sim \begin{bmatrix} a & b & c & d & e & f & g & h & i & j & k & l \\ f & a & b & c & d & e & l & g & h & i & j & k \\ e & f & a & b & c & d & k & l & g & h & i & j \\ d & e & f & a & b & c & j & k & l & g & h & i \\ c & d & e & f & a & b & i & j & k & l & g & h \\ b & c & d & e & f & a & h & i & j & k & l & g \\ g & h & i & j & k & l & a & b & c & d & e & f \\ l & g & h & i & j & k & f & a & b & c & d & e \\ k & l & g & h & i & j & e & f & a & b & c & d \\ j & k & l & g & h & i & d & e & f & a & b & c \\ i & j & k & l & g & h & c & d & e & f & a & b \\ h & i & j & k & l & g & b & c & d & e & f & a \end{bmatrix} \qquad (31.1)$$

The 12-dimensional algebras within this group, with the Standard form Cayley table above, (31.1) are of the form:

$$a + b\sqrt[6]{\pm 1} + c\sqrt[3]{\pm 1} + d\sqrt[2]{\pm 1} + e\sqrt[3]{\pm 1} + f\sqrt[6]{\pm 1} + g\sqrt[2]{\pm 1} + h\sqrt[6]{\pm 1} + i\sqrt[6]{\pm 1} + j\sqrt[2]{\pm 1} + k\sqrt[6]{\pm 1} + l\sqrt[6]{\pm 1} \quad (31.2)$$

Not manifest in our space-time:

The presence of the cube roots and the 6th roots means that this algebra can never be manifest in our 4-dimensional space-time because our 4-dimensional space-time holds only 2-dimensional spinor algebras and these 2-dimensional spinor algebras have only square roots. Thus this algebra is not part of the physics of our universe.

The algebras:

The parameter elimination code of the order twelve $C_2 \times C_6$ group leaves eleven free parameters. The code ends with a single quadratic elimination equation. As with the $C_2 \times C_2 \times ...$ algebras, it is the negative root of this quadratic elimination equation that leads to the non-commutative algebras.

When we take the positive root of the quadratic parameter elimination equation, we get six non-isomorphic commutative algebras:

$$128 \quad \textit{off} \quad 1+6\sqrt[6]{+1}+2\sqrt[3]{-1}+3\sqrt[2]{+1} \quad \textit{Comm} \tag{31.3}$$

$$128 \quad \textit{off} \quad 1+6\sqrt[6]{+1}+2\sqrt[3]{+1}+3\sqrt[2]{+1} \quad \textit{Comm} \tag{31.4}$$

$$256 \quad \textit{off} \quad 1+6\sqrt[6]{+1}+\sqrt[3]{+1}+3\sqrt[2]{+1}+\sqrt[3]{-1} \quad \textit{Comm} \tag{31.5}$$

$$384 \quad \textit{off} \quad 1+2\sqrt[6]{+1}+2\sqrt[3]{+1}+\sqrt[2]{+1}+2\sqrt[6]{-1}+4\sqrt[6]{-1} \quad \textit{Comm} \tag{31.6}$$

$$384 \quad \textit{off} \quad 1+4\sqrt[6]{-1}+2\sqrt[3]{-1}+2\sqrt[2]{-1}+\sqrt[2]{+1}+2\sqrt[6]{+1} \quad \textit{Comm} \tag{31.7}$$

$$768 \quad \textit{off} \quad 1+2\sqrt[6]{+1}+\sqrt[3]{+1}+\sqrt[2]{+1}+\sqrt[3]{-1}+2\sqrt[2]{-1}+4\sqrt[6]{-1} \quad \textit{Comm} \tag{31.8}$$

When we take the negative root of the quadratic parameter elimination equation, we get six non-isomorphic non-commutative algebras:

$$128 \quad \textit{off} \quad 1+6\sqrt[6]{-1}+2\sqrt[3]{-1}+3\sqrt[2]{-1} \quad \textit{Non-Comm} \tag{31.9}$$

$$128 \quad \textit{off} \quad 1+6\sqrt[6]{-1}+2\sqrt[3]{+1}+3\sqrt[2]{-1} \quad \textit{Non-Comm} \tag{31.10}$$

$$256 \quad \textit{off} \quad 1+6\sqrt[6]{-1}+\sqrt[3]{-1}+3\sqrt[2]{-1}+\sqrt[3]{+1} \quad \textit{Non-Comm} \tag{31.11}$$

$$384 \quad \textit{off} \quad 1+4\sqrt[6]{+1}+2\sqrt[3]{+1}+2\sqrt[2]{+1}+2\sqrt[6]{-1}+\sqrt[2]{-1} \quad \textit{Non-Comm} \tag{31.12}$$

$$384 \quad \textit{off} \quad 1+2\sqrt[6]{-1}+2\sqrt[3]{-1}+\sqrt[2]{-1}+2\sqrt[2]{+1}+4\sqrt[6]{+1} \quad \textit{Non-Comm} \tag{31.13}$$

$$768 \quad \textit{off} \quad 1+4\sqrt[6]{+1}+\sqrt[3]{+1}+2\sqrt[2]{+1}+\sqrt[3]{-1}+2\sqrt[6]{-1}+\sqrt[2]{-1} \quad \textit{Non-Comm} \tag{31.14}$$

We see that, like the $C_2 \times C_2$ group, the quadratic parameter elimination equation has led to a set of non-commutative algebras.

Chiral algebras:

The non-commutative algebras of the $C_2 \times C_6$ group are chiral algebras; by this we mean that the non-commutative pairs of variables have a zero anti-commutator.

Commutation relations:

We form the commutation table of the $C_2 \times C_6$ algebras by copying the Standard Form Cayley table above, (31.1) and putting zeros in place where the commutator is zero. For example, the top left-hand 6×6 corner is a copy of the commutative C_6 group and all these variables commute with each other. This means we have zeros in the top left-hand 6×6 corner and the bottom right-hand 6×6 corner of the commutation table. The variables $\{c, e\}$ commute with every other variable, and so the c column is all zeros as is the c row, and the e column is all zeros as is the e row.

Other pairs of variables commute with each other; for example, the $\{g, k\}$ variables commute with each other. Of course, a variable commutes with itself.

We have the commutation table:

$$\text{Commutation Table}_{C_2 \times C_6} \sim$$

	a	b	c	d	e	f	g	h	i	j	k	l
a	0	0	0	0	0	0	0	0	0	0	0	0
f	0	0	0	0	0	0	±l	±g	±h	±i	±j	±k
e	0	0	0	0	0	0	0	0	0	0	0	0
d	0	0	0	0	0	0	±j	±k	±l	±g	±h	±i
c	0	0	0	0	0	0	0	0	0	0	0	0
b	0	0	0	0	0	0	±h	±i	±j	±k	±l	±g
g	0	±h	0	±j	0	±l	0	0	0	0	0	0
l	0	±g	0	±i	0	±k	0	0	0	0	0	0
k	0	±l	0	±h	0	±j	0	0	0	0	0	0
j	0	±k	0	±g	0	±i	0	0	0	0	0	0
i	0	±j	0	±l	0	±h	0	0	0	0	0	0
h	0	±i	0	±k	0	±g	0	0	0	0	0	0

(31.15)

Unlike the quaternions, there are only a few instances of variables which are non-commutative. These commutation relations are all chiral.

In general:

The distribution of the chiral algebras throughout the finite groups is not clearly understood. It seems reasonable to expect that every finite group with a $C_2 \times C_2$ sub-group or a $C_2 \times C_2 \times C_2$ subgroup would have some chiral algebras as in the case of this $C_2 \times C_6$ group, but the 16-dimensional $C_2 \times C_2 \times C_2 \times C_2$ group has no chiral algebras even though it has $C_2 \times C_2$ sub-groups and non-commutative algebras.

Chapter 32

Concluding Remarks

We have found the 4-dimensional chiral algebras. We have a reasonably clear understanding of 4-dimensional chirality, and it seems to be exactly what we find in the particle physics of electrons and neutrinos. We are very happy with this, and we take the view that the Lie algebra $SU(2)$ used in particle physics should be replaced by the quaternion algebras.

We have found the 8-dimensional chiral algebras. We do not have a reasonably clear understanding of 8-dimensional chirality, but it seems that it might be connected to the particle physics of quarks. We need a deeper understanding of the 8-dimensional algebras before we advocate replacing the Lie algebra $SU(3)$ with the 8-dimensional chiral algebras.

We have discovered that there are no 16-dimensional chiral algebras. Although not chiral, the 16-dimensional algebras are not well understood; there is much work to be done here.

We have briefly looked at the chiral algebras of other groups even though these chiral algebras cannot be manifest in our 4-dimensional space-time.

We would have liked to have given the reader a clear and simple understanding of the 8-dimensional chiral algebras as we have done with the 4-dimensional chiral algebras. We have failed in this regard. None-the-less, we have clearly demonstrated that mathematics is left-handed.

Dennis Morris

Brotton

March 2017

Other Books by the Same Author

The Naked Spinor – a Rewrite of Clifford Algebra

Spinors exist in Clifford algebras. In this book, we explore the nature of spinors. This book is an excellent introduction to Clifford algebra.

Complex Numbers The Higher Dimensional Forms – Spinor Algebra

In this book, we explore the higher dimensional forms of complex numbers. These higher dimensional forms are connected very closely to spinors.

Upon General Relativity

In this book, we see how 4-dimensional space-time, gravity, and electromagnetism emerge from the spinor algebras. This is an excellent and easy-paced introduction to general relativity.

From Where Comes the Universe

This is a guide for the lay-person to the physics of empty space.

Empty Space is Amazing Stuff – The Special Theory of Relativity

This book deduces the theory of special relativity from the finite groups. It gives a unique insight into the nature of the 2-dimensional space-time of special relativity.

The Nuts and Bolts of Quantum Mechanics

This is a gentle introduction to quantum mechanics for undergraduates.

Quaternions

This book pulls together the often separate properties of the quaternions. Non-commutative differentiation is covered as is non-commutative rotation and non-commutative inner products along with the quaternion trigonometric functions.

The Uniqueness of our Space-time

This book reports the finding that the only two geometric spaces within the finite groups are the two spaces that together form our universe. This is a startling finding. The nature of geometric space is explained alongside the nature of division algebra space, spinor space. This book is a catalogue of the higher dimensional complex numbers up to dimension fifteen.

Lie Groups and Lie Algebras

This book presents Lie theory from a diametrically different perspective to the usual presentation. This makes the subject much more intuitively obvious and easier to learn. Included is perhaps the clearest and simplest presentation of the true nature of the Lie group $SU(2)$ ever presented.

The Physics of Empty Space

This book presents a comprehensive understanding of empty space. The presence of 2-dimensional rotations in our 4-dimensional space-time is explained. Also included is a very gentle introduction to non-commutative differentiation. Classical electromagetism is deduced from the quaternions.

The Electron

This book presents the quantum field theory view of the electron and the neutrino. This view is radically different from the classical view of the electron presented in most schools and colleges. This book gives a very clear exposition of the Dirac equation including the quaternion rewrite of the Dirac equation. This is an excellent introduction to particle physics for students prior to university, during university and after university courses in physics.

The Quaternion Dirac Equation

This small book (only 40 pages) presents the quaternion form of the Dirac equation. The neutrino mass problem is solved and we gain an explanation of why neutrinos are left-chiral. Much of the material in this book is drawn from 'The Electron'; this material is presented concisely and inexpensively for students already familiar with QFT.

An Essay on the Nature of Space-time

This small and inexpensive volume presents a view of the nature of empty space without the detailed mathematics. The expanding universe and dark energy is discussed.

Elementary Calculus from an Advanced Standpoint

This book rewrite the calculus of the complex numbers in a way that covers all division algebras and makes all continuous complex functions differentiable and integrable. Non-commutative differentiation is covered. Gauge covariant differentiation is covered as is the covariant derivative of general relativity.

Even Mathematicians and Physicists make Mistakes

This book points out what seems to be several important errors of modern physics and modern mathematics. Errors like the misunderstanding of rotation, the failure to teach the higher dimensional complex numbers in most universities, and the mathematical inconsistency of the Dirac equation and some casual errors are discussed. These errors are set in their historical circumstances and there is discussion about why they happened and the consequences of their happening. There is also an interesting chapter on the nature of mathematical proof within our society, and several famous proofs are discussed (without the details).

Finite Groups – A Simple Introduction

This book introduces the reader to finite group theory. Many introductory books on finite groups bury the reader in geometrical examples or in other types of groups and lose the central nature of a finite group. This book sticks firmly with the permutation nature of finite groups and elucidates that nature by the extensive use of permutation matrices. Permutation matrices simplify the subject considerably. This book is probably unique in its use of permutation matrices and therefore unique in its simplicity.

Non-commutative Differentiation and the Commutator

(The Search for the Fermion Content of the Universe)

This book develops the theory of non-commutative differentiation from the fundamentals of algebra. We see what an algebraic operation (addition, multiplication) really is, and we discover that the commutator is a third fundamental algebraic operation within some division algebras. This leads to the first part of the derivation of the fermion content of the universe.

Index

B

C

D

E

F

G